Tiberius Cavallo

An Essay on the Medicinal Properties of Factitious Airs

With an appendix, on the nature of blood

Tiberius Cavallo

An Essay on the Medicinal Properties of Factitious Airs
With an appendix, on the nature of blood

ISBN/EAN: 9783337393144

Printed in Europe, USA, Canada, Australia, Japan

Cover: Foto ©berggeist007 / pixelio.de

More available books at **www.hansebooks.com**

AN

ESSAY

ON THE

MEDICINAL PROPERTIES

OF

FACTITIOUS AIRS.

WITH AN

APPENDIX,

ON THE NATURE OF BLOOD.

BY

TIBERIUS CAVALLO, F.R.S.

LONDON:

PRINTED FOR THE AUTHOR,
And fold by C. DILLY, in the Poultry; P. ELMSLY,
and D. BREMNER, in the Strand.—1798.

IT is not quite forty years fince the artificial aerial fluids began to be adminiftered as remedies to the human body. The uncertainty, and the errors of the early applications, rendered the progrefs of the practice flow and doubtful; nor has the experience, or the fuccefs of recent and more numerous practitioners, been fufficient to determine the precife power of the aerial fluids, or to diffipate the doubts which are ftill entertained concerning their ufe.

The defire of extricating the fubject from the conflict of contrary opinions, eftablifhed prejudices, and oppofite interefts, has induced the Author, perhaps too haftily, to publifh the prefent work, which, in every fenfe of the word, deferves the epithet of imperfect. But he hopes that the importance of an object fo highly interefting teresting to the human species, may palliate, if not justify, the imperfections of the performance, which might, perhaps, have been less excusable in other subjects.

To exhibit a concise view of ascertained facts, to separate them from suppositions and hypotheses, and to point out the ways of investigating the farther use of factitious airs, has been the Author's principal aim in the compilation of the present Essay.

In the course of his inquiries, he has frequently found cause to admire the ingenuity, the caution, and the perseverance of several gentlemen, who either have administered the aerial fluids, or have otherwise exerted themselves in the promotion of their use. Yet he has taken particular care to avoid paying them any compliments, or even making frequent use of their names, lest his desire of promoting the subject should be apparently converted into an endeavour of promoting the interest of certain practitioners.

The first four chapters contain such facts as may be of theoretical use in the applications

PREFACE. v

tions of aeriform fluids, and in the inveftigation of their action, independent on medical cafes. The fifth chapter exhibits a concife view of the modern theory of aerial fluids, and of the proceffes that are principally depending thereon, fuch as refpiration, combuftion, &c. The fixth and feventh chapters fhew the practical application of thofe fluids by way of remedies to the human body; and this practice is exemplified in the eighth chapter, in which a felect number of authentic cafes is related. The ninth, or laft chapter, contains feveral practical remarks, hints, &c. which could not be conveniently inferted in the preceding part of the work.

Laftly, a differtation on the nature and properties of blood has been added by way of Appendix, that fluid being evidently and principally concerned in refpiration, and in the general dependance of the animal exiftence on the aerial fluids.

By the mixt ufe of the old and the new chemical names in various parts of the work, the author imagines that his meaning

may

PREFACE.

may be rendered lefs equivocal, and more generally intelligible; for at a time when the old names are not quite difufed, and the new chemical nomenclature not univerfally underftood, it is difficult to determine whether the greateft number of readers may remain fatisfied with the exclufive ufe of either.

Wells Street,
January the 8th, 1798.

CONTENTS.

CHAP. I. *THE principal Properties of thofe Airs, or permanently elaſtic Fluids, which have been applied as Remedies to the Human Body* - - page 1

II. *Facts concerning the Refpiration of Common, and of Oxygen, Airs* - - - 22

III. *Phænomena arifing from breathing other Aerial Fluids, befides the Common and the Oxygen Airs* - - - 40

IV. *Phænomena arifing from the Application of the above-mentioned elaſtic Fluids to other Parts of the Animal Body befides the Lungs* - - - 51

V. *Theory of the Nature of Aerial Fluids, and of Refpiration* 58

CHAP.

CONTENTS.

CHAP. VI. *A general Idea of the Application of Aerial Fluids for the Cure of Diforders incident to the Human Body* - page 88

VII. *Of the particular Adminiſtration of Aerial Fluids in different Diforders* - - 114

VIII. *Medical Cafes in which Aerial Fluids were adminiſtered* 149

IX. *Practical Remarks, Hints, &c.* 202

APPENDIX.

On the Nature of Blood - - - 217

AN
ESSAY
ON THE
MEDICINAL PROPERTIES
OF
FACTITIOUS AIRS.

CHAPTER I.

The principal Properties of thofe AIRS, *or permanently elaftic Fluids, which have been applied as Remedies to the Human Body.*

THE philofophical inveftigations of the two laft centuries, and particularly of the prefent age, have afcertained the exiftence of various elaftic fluids, analogous to common air, with refpect to elafticity and invifibility; but otherwife effentially different from it, as alfo different from each other; fuch are the *dephlogifticated air*, or *vital air*, or *oxygen air*; the *phlogifticated air*, or *gas azote*; the *fixed air*, or *carbonic acid gas*;

2 MEDICINAL PROPERTIES *of*

gas; the *inflammable air*, or *hydrogen gas*; the *nitrous gas*, &c. But as of all the different airs five only appear to be applicable to the human body, *viz.* the common, the oxygen, the azotic, the carbonic acid, and the hydrogen airs, we shall not therefore extend our notice to any other sort of elastic fluid; nor shall we describe more than the principal properties of those five; *viz.* such properties only as may be useful to elucidate their action on the human body.

Of the Common, or Atmospherical Air.

THAT invisible elastic fluid, which surrounds the earth, and in which we live, is indispensably necessary to animal life, to combustion, and to other processes. No animal can live, nor can any combustible body burn, without air. For either purpose the atmospherical air is more or less useful in proportion to its purity.

WHEN common air is mixed with another particular sort of air, called *nitrous gas*, a diminution of bulk takes place, which is proportionate to the purity of the air; the

*   purest

purest air being diminished most, and *vice versa*; so that very impure air suffers no diminution. Hence the quality or goodness of common air may be ascertained by mixing a certain quantity of it with a determinate quantity of nitrous air, and then measuring the diminution of bulk that ensues. The instrument in which this operation for ascertaining the purity of the air is made, has been called an *eudiometer*.

The purity of common air is not the same in all places, nor is it constant in the same place at all times. The variation in the latter case is much more considerable than in the former; yet, upon the whole, it is not very great. If in the usual state of the atmosphere, and in places that are reckoned healthy, 100 parts or measures of common air be mixed with an equal quantity of nitrous air, their bulk, after the mixture, will be found, instead of 200 parts, to be between 100 and 120, more or less, according to the time of the year, situation of the place, state of the atmosphere, &c.

But in caves, mines, crowded rooms, hofpitals, work-fhops, and the like, the air is lefs pure; yet even in this cafe the difference, as indicated by the teft of nitrous air, is but trifling; excepting indeed thofe places in which the communication with the external air is abfolutely or almoft entirely interrupted *.

NOTWITHSTANDING the fmall difference which is manifefted by this method of trying the purity of common air, it is however evident, from the oppreffion which is felt in certain inftances, and the reviving effect which is experienced in other cafes,

* Dr. Prieftley having dined one day in company with eight or ten perfons, in a large and very lofty room, and happening to go out of the room for a fhort time, was, on his return, ftruck with the offenfivenefs of the air, and his curiofity prompted him to afcertain the degree in which the air was injured. On trial he found that 100 parts of that air, with 100 parts of nitrous air, were reduced to 131 parts; whereas the like experiment being performed with the air of a well-ventilated room of the fame houfe, the 200 parts of mixt aerial fluid were reduced to 125 parts.

that

that the human lungs are fenfibly affected by the fmalleft differences in the purity of the air. But it is neceffary to remark, that noxious particles are frequently fufpended in common air, which do not alter the effect of nitrous gas upon it, though, at the fame time, they render it very offenfive to animals.

CONSIDERING the variety of vapours, minute bodies, &c. that are continually fcattered through, and float in, the air, the atmofphere muft be looked upon as being always contaminated by the prefence of minute animal, vegetable, and even mineral, particles;—of bodies, in fhort, that are foreign to, or unconnected with, the nature of air.

THE quality of common air is not altered by merely heating or cooling *, or by

* Every degree of Fahrenheit's thermometer rarifies or increafes the bulk of common air, by about $\frac{1}{475}$ part of the whole.

keeping, or by being for a time loaded with the vapour of water, nor by rarefaction or condenfation; but it is contaminated principally by refpiration, by combuftion, by the fermentation and putrefaction of animal and vegetable bodies, by the calcination of metallic fubftances, by the prefence of vegetables when they are not under the influence of the fun's rays, and by the admixture of every other gas, or permanently elaftic fluid, except the oxygen.

WHEN the common air is completely contaminated, or rendered unfit for combuftion and refpiration, it is (according to the prefent nomenclature) called *gas azote*, whereas it was formerly called *phlogifticated air*.

VITIATED air is capable of being meliorated various ways, and the methods of effecting it may be diftinguifhed into natural and artificial. The natural means are far from being known to their full extent; but the vegetation of plants, in certain circumftances, cumſtances, and the contact of water, as in rains, dews, &c. are two very powerful correctors of contaminated air. Whether thoſe and other natural means, are ſufficient to preſerve the atmoſpherical air nearly in the ſame degree of purity, or whether that degree be conſtantly undergoing a gradual change, ſo as to render the air either continually better or worſe, is a very intereſting queſtion, but it can only be anſwered by the philoſophers of future generations. For my part, I am led to ſuſpect that the purity of the air is ſubject to a periodical fluctuation, or to an alternate increaſe and decreaſe for an uncertain number of years.

VENTILATION, and whatever promotes ventilation, does nothing more than remove vitiated air from thoſe places in which it is generated, and diſperſe it through the atmoſphere.

THE artificial methods of correcting vitiated air are few and imperfect. Ventilation, by means of bellows and other machines,

chines, is the moſt efficacious, and at the same time the moſt practicable way of improving the air of hoſpitals, ſick rooms, priſons, &c. *viz.* by removing the vitiated, and introducing a freſh current of purer air. A fire purifies the air of certain places, only by promoting the ventilation or circulation, and by drying the moiſture; but the air which has paſſed through the fire muſt not remain in thoſe places, otherwiſe the injury will be infinitely greater than the advantage. It has been confidently aſſerted, and denied, but it is now with limitation believed, that the vapours of nitrous, or of marine acid, will diveſt common air of the poiſonous effluvia of contagious diſorders; hence the vapours of thoſe acids are now frequently diſperſed through the air of hoſpitals, crowded ſhips, &c. When noxious vapours are merely ſuſpended in the air, as it often takes place in ſeveral natural and artificial proceſſes, then reſt alone, or at moſt a ſlight agitation in water will be ſufficient to purify the air. By the admixture of oxygen gas, a quantity of common air may be improved

proved to almoſt any degree; but the method is difficult and expenſive; hence it can only be uſed with limitation in certain caſes, which will be ſpecified in the ſequel.

Of the Dephlogiſticated, or Oxygen Air.

THE oxygen is a ſort of aerial fluid, that poſſeſſes the uſeful properties of common air in a much more eminent degree; *viz.* it aſſiſts combuſtion and animal reſpiration for a much longer time, and with ſuperior energy. When a lighted candle is introduced into a veſſel full of oxygen air, its flame becomes larger, and ſurpriſingly brighter than in common air. Its heat is likewiſe increaſed to a very great degree.

THIS air is not found pure or unmixed in nature, but it may be extracted from various ſubſtances by means of artificial proceſſes. The leaves of plants, indeed, yield a conſiderable quantity of it whilſt they are expoſed to the light of the ſun; but the oxygen air which is thus produced, mixes with,

with, and is difperfed through the circumambient air as foon as it is generated; fo that the air contiguous to the plants is feldom fenfibly better than that of the neighbouring country.

By the addition of nitrous air the oxygen is diminifhed much more than common air. When 100 parts of good oxygen air are mixed with an equal quantity, *viz.* 100 parts, of nitrous air, their joined bulk will not exceed 50 parts, the other 150 parts having loft the aerial form. Nor is this the utmoft degree of diminution that can be produced; for if 100 parts of the pureft oxygen that can be procured, be mixed with twice its quantity of nitrous gas, almoft the whole bulk of elaftic fluid will difappear; at moft, the refiduum will not exceed five or fix parts. By putting a lighted candle into a veffel full of any fpecies of refpirable air, and obferving the effect of that air on the flame, one may eftimate the degree of its purity near enough for feveral purpofes.

THE

The following are the principal methods of procuring this air. The green leaves of vegetables, when placed in a glafs receiver full of, and inverted in fpring water, and thus expofed to the direct rays of the fun, yield a confiderable quantity of oxygen air, which afcends to the upper part of the receiver, and may be eafily removed from it for ufe. One hundred leaves of Indian crefs, *nafturtium Indicum*, in a gallon of fpring water, will, in about three hours expofure to the fun, yield about ten cubic inches of oxygen air, not indeed quite pure, but yet vaftly better than common air. I do not know of any plant whofe leaves produce this fort of air in greater abundance.

There are feveral fubftances from which oxygen air may be extracted by the action of heat or of acids; but thofe which upon the whole yield it in greateft plenty, and are fit to be ufed, are faltpetre or nitre, and the metallic calces.

ONE ounce of nitre, by remaining expofed to a full red, or rather a white, heat in an earthen retort for about four or five hours, will give between 700 and 800 cubic inches of oxygen air, which is not equally good in every period of the procefs, but at a medium it is fuch that if 100 parts of it be mixed with 150 parts of nitrous air, the whole will be reduced to about 100 parts. This oxygen gas contains a quantity of nitrous acid in the form of vapour, and therefore, when it is to be ufed for refpiration, the acid vapour muft be previoufly feparated from it, which may be done by agitating the air in an alkaline lixivium, or at leaft in lime water.

IF an ounce of *mercurius precipitatus per fe* be expofed to a barely red heat in a glafs veffel, it will yield at leaft 66 cubic inches of very good oxygen air.

RED precipitate of mercury, when treated in the like manner, does alfo yield a confiderable quantity of this fort of air.

THE action of a red heat alone, or of vitriolic acid and a moderate degree of heat, expels from minium, or red lead, about ten or twelve times its bulk of oxygen, mixed with about one third of carbonic acid, air; the latter of which may be feparated from the former by wafhing in lime water. If the red lead be previoufly moiftened with nitrous acid, and then ftrong vitriolic acid be poured upon it, a greater quantity of oxygen gas will be obtained in a fhorter time, and even without the application of heat.

THIS fort of elaftic fluid may be alfo obtained in fmall quantities from feveral other metallic calces; but the mineral called *manganefe*, gives a great quantity of it in an eafy manner; it is at the fame time a very cheap article, fo that, upon the whole, manganefe is at prefent the moft eligible fubftance for the purpofe of procuring oxygen air.

MANGANESE

Manganese is not always of the fame quality, and of courfe the elaftic fluid, which is extracted from a given quantity of it, is variable both in quantity and quality. One ounce of good manganefe, free from large calcareous particles, will, in a red heat, yield more than two pints and a half wine meafure, or about eighty cubic inches of elaftic fluid, about one tenth of which is carbonic acid, and the reft is oxygen gas. By means of vitriolic acid and a gentle heat, about an equal quantity of elaftic fluid, nearly of the fame quality, may be extracted from manganefe; but in this cafe fome acid vapours come over with it, which muft be carefully wafhed off in order to render the oxygen air fit for refpiration.

The oxygen air is diminifhed to a much greater degree than common air, not only by the admixture of nitrous gas, but alfo by all the proceffes which are known to diminifh atmofpherical air ; and indeed fometimes the whole quantity of oxygen air is abforbed or deprived of its aerial form.

form. Thus, by respiration, this air will be entirely absorbed, excepting indeed that part which is converted into fixed air.

Of Fixed Air, or the Carbonic Acid Gas.

THIS gas, which is the heaviest of the aerial fluids, is of an acid nature, but it reddens only light blue vegetable colours; it crystallizes with fixed alkali, and is possessed of a considerable antiseptic power. It is absolutely incapable of assisting respiration and combustion *; nor is it diminished by nitrous air. It combines with various substances, and is readily absorbed by water, to which it communicates an acidolous taste and sparkling property. It is also absorbed by, and precipitates the calcareous earth in lime water, but when in greater quantity, it again dissolves the cal-

* Even a mixture of one part of fixed, and eight parts of common, air will extinguish the flame of a candle. See Cavendish's paper, in the Phil. Transf. for 1766.

careous earth in the water. It alſo diſſolves iron in water, and keeps it diſſolved therein.

THIS elaſtic fluid is produced in a great many natural as well as artificial proceſſes. It is frequently found in ſubterranean places, eſpecially in the vicinity of volcanos, and hot ſprings, where, on account of its great ſpecific gravity, it remains for a conſiderable time, unleſs it be removed by means of ventilation, &c. It is contained more or leſs in almoſt all the mineral waters; it is abundantly produced in vinous fermentation. Reſpiration, combuſtion, and ſome other proceſſes, do likewiſe produce a certain quantity of carbonic acid gas. It is contained in a variety of mineral ſubſtances, and particularly in calcareous earth, as chalk, marble, &c. from which ſubſtances a great quantity of that gas may be extracted by means of heat or of acids *; the calcareous

---

* The fixed air which is contained in white marble amounts to about one third part of its weight.

bodies

bodies remaining, after the lofs of that gas, in a cauftic or acrid ftate; fo that the calcareous earth, by being in a mild ftate whilft it contains that elaftic fluid, may be juftly confidered as a neutral falt, confifting of an earthy bafis and an aerial acid.

Of the Inflammable Air, or Hydrogen Gas.

INFLAMMABLE Air is the lighteft of the elaftic fluids. It is, as its name imports, a combuftible fluid, which, like other combuftible fubftances, may be inflamed by the contact of an ignited body, and will burn only when in contact with common, or oxygen, air.

THOUGH this fort of elaftic fluid be abfolutely unfit for refpiration, it is not, however, fo noxious as the carbonic acid. It fuffers no diminution when mixed with nitrous air. Its bulk is increafed of $\frac{1}{400}$ part of the whole by each degree of Fahrenheit's thermometer.

C HYDROGEN

Hydrogen gas is abundantly produced during the diffolution of animal and vegetable bodies; hence it is often found to come out of ponds, burying grounds, and other places that contain animal and vegetable matter in a ftate of decay. This gas does alfo frequently come out of the earth, where inflammable minerals are contained, as in coal mines, and mines of fulphureous metallic ores. But in all thofe cafes the inflammable gas, by being much lighter than common air, afcends to the upper regions of the atmofphere as foon as it is produced, and leaves the air, adjacent to the ground, very little, if at all, infected, excepting in vaulted fubterranean places, where, indeed, befides its infecting the common air, it fometimes takes fire and explodes, to the great danger of the miners.

By means of heat, or of acids, this gas may be obtained from almoft all forts of bodies, whether they be vegetable, animal, or mineral. But the greateft quantity of it may be extracted from iron, or from zinc, by means of diluted vitriolic acid; and likewife from iron, by paffing the fteam of boiling water over its furface, the iron being red hot. When charcoal is treated in the laft-mentioned manner, it likewife yields abundance of a peculiar fort of inflammable gas, called *hydrocarbonate*, which however is mixed with a confiderable proportion of carbonic acid gas.

HYDROGEN gas has the property of diffolving and holding in fufpenfion, for a longer or fhorter time, a variety of fubftances, fuch as iron, charcoal, fulphur, phofphorus, &c. from which circumftance it acquires a variety of particular names as well as properties. Hence we hear of the *phofphoric hydrogen gas*, or *phofphuret* of *hydrogen*; of the *fulphuric hydrogen gas*, or *fulphuret* of *hydrogen*, &c.; hence alfo we find that the hydrogen gas is not always of the fame fpecific gravity, nor has it always the fame fmell.

It has been obferved, that the hydrogen gas fometimes lofes its inflammability, and degenerates into azotic air. This change happens more frequently when the hydrogen gas is mixed with common air. The caufe of this phenomenon has not yet been fully afcertained.

For the fake both of brevity and of perfpicuity I have omitted to mention the fpecific gravities of the abovementioned elaftic fluids in the preceding pages, and fhall add them all together in the following table, which contains their fpecific gravities as well as the abfolute weight of a cubic inch of each elaftic fluid.

The gravity of common air is confiderably affected by the variations of heat, wind, purity, &c. fo that its fpecific gravity, compared with that of water, has fometimes been known to be as one to fix hundred and fix, and at other times as one to nine hundred and thirty-one[*]. The gra-

[*] Muffchenbroek, tom. II. §. 2059.

FACTITIOUS AIRS. 21

vities of other elaſtic fluids are likewiſe ſubject to the ſame variations. But the following table has been calculated for a mean and temperate ſtate of the air, viz. when its gravity is to that of water, as one to eight hundred, when the height of the barometer is 29,85 inches, and when Fahrenheit's thermometer is at 55°.

Names of the elaſtic Fluids.	Their ſpecific Gravities.	Abſolute Weight of a Cubic Inch of each in Troy Grains.
Common air	1	0,31648
Azotic gas, or common air completely diminiſhed by nitrous gas	0,948	0,3
Oxygen air	1,0427	0,33
Carbonic acid gas	1,5	0,475
The lighteſt hydrogen gas	0,0833	0,02637

CHAPTER II.

Facts concerning the Respiration of Common, and of Oxygen, Air.

THAT the whole mass of air which surrounds the earth is called the atmosphere, that this atmosphere extends to a considerable but unknown distance above the surface of the earth, that it decreases in density as it recedes from the earth, that its motion is called wind, that it acts upon all other bodies by its temperature, its weight, and other qualities, that it absorbs vapours, or keeps them suspended, and such other like properties of the atmospherical fluid, have been rendered so common by the present state of knowledge and of polite education, as not to demand any particular elucidation in this work; we shall, therefore, proceed immediately to enumerate the phenomena which have been ascertained relatively to the respiration of common air,

upon

FACTITIOUS AIRS. 23

upon which, as upon a folid bafis, we may afterwards eftablifh the theory and the practice of applying the factitious airs to the human lungs.

A CERTAIN quantity of air will fupport animal life, or combuftion, but for a limited time. If a lighted middle-fized tallow candle be confined in a veffel that holds one gallon of common air, the flame will, in a few feconds of time, begin to grow dim, and it will be extinguifhed at the end of about one minute; after this, if another lighted candle be introduced into the fame veffel, its flame will be extinguifhed immediately.

IF a man be confined in a veffel that holds ten gallons of common air, he will begin to feel an oppreffion, and a difficulty of refpiration, at the end of eight or ten minutes; this difficulty will gradually increafe, and at the end of about half an hour, reckoning from the beginning of his confinement, he will lofe his fenfation, his motion, and, prefently after, his life. The fame effect will take place with other animals, in a longer or fhorter time, proportionably to their fize, nature, and difpofition of body.

In the ufual way of breathing, when refpiration is performed in a natural and eafy manner, a full grown perfon confumes about five cubic feet, or thirty gallons and a half, beer meafure, of common air per hour.

A MAN generally performs one infpiration and one expiration for every feven or eight pulfations of his arteries; therefore reckoning, at a mean, eighty pulfations per minute, a perfon may be faid to perform eleven or twelve infpirations, and as many expirations, in a minute. But refpiration is quickened by various caufes; *viz.* by the quickening of the pulfe, by agitation of the body, by heat, by furprife, by difeafes of the lungs, by a rarefied atmofphere, and by impure air. Thus when a man is confined

in

FACTITIOUS AIRS. 25

in a certain quantity of air, his refpiration is quickened in proportion as that quantity of air becomes contaminated; he alfo takes in and expels a greater quantity of air at a time, in order to compenfate for the want of purity. The fame quickening of refpiration takes place on high mountains, where the air is more rare than on the level of the fea.

At a medium, about 30 cubic inches of air are taken in at one infpiration, and a quantity, nearly equal to it, is thrown out at every expiration; but a great deal of air remains in the lungs, wind-pipe, and mouth; fo that by a violent expiration after a natural infpiration, a double quantity, *viz.* fixty cubic inches of air, may be expelled, and even then fome air neceffarily remains in the lungs, wind-pipe, and mouth.

The air which has ferved for one infpiration is not thereby completely contaminated, but it may be refpired again and again. 350 cubic inches of common air
were

were confined in a bladder that was furnished with a wooden tube; this tube was applied to the mouth of a healthy middle-aged man, who, ſtopping his noſtrils, endeavoured to breathe that quantity of air as long as he poſſibly could. After having performed forty inſpirations, his ſtrength began to fail, and he was obliged to deſiſt.

OLD perſons, people of a bad habit of body, or labouring under diſeaſes, and ſuch as eat and drink immoderately, will contaminate the air much faſter than the healthy, the moderate, and the young.

IT has been aſſerted, that ſome human beings can live with a much ſmaller quantity of air than has been mentioned above, and that divers have ſometimes been known to remain under water ten or fifteen minutes, and even a longer time*. It has

---

* See Beckman's Hiſtory of Inventions, article *Diving-Bell*; and Gmelin's Reiſe Durch Ruſsland, II. p. 199.

been

been likewife difcuffed, whether fuch divers were enabled to remain fo long under water, and without air, by any particular conformation of the internal parts of their bodies, or from long practice and particular artifices. But there are ftrong reafons for difcrediting the above-mentioned affertions. The inaccurate way of reckoning the time in fuch cafes, and the common fondnefs for the marvellous, are in general the foundation of fuch extraordinary reports. Upon the whole, it will be found, that the moft experienced diver can hardly remain without air longer than a minute and a half; but moft perfons will begin to feel a degree of uneafinefs in about half a minute's time.

THE air, which has been completely contaminated by refpiration, is deleterious to other animals, though fmall and young animals will live a fhort time in it: it extinguifhes flame, is diminifhed very little by nitrous air, contains about one-thirtieth of carbonic acid gas, and is contracted in bulk,

bulk, the diminution being various, but hardly ever exceeding one-fifth part of the original quantity.

THE deleterious quality of the air that has been contaminated by refpiration is in great meafure owing to the carbonic acid gas, which is formed in the procefs of refpiration; and it is for this reafon that, when an animal is confined in a veffel full of refpirable air, he will be able to live longer in it when fome lime-water is placed in the veffel, than otherwife; becaufe the lime-water abforbs the carbonic acid gas as foon as it is generated. An animal will likewife live longer in a veffel full of air, when he is placed at the upper than at the lower part of the veffel; becaufe in the former cafe the carbonic acid gas will, on account of its great fpecific gravity, fall towards the lower part of the veffel, and will, of courfe, be at a diftance from the body of the animal.

THE refpiration of oxygen air is attended with peculiar phenomena. The oxygen, like the common, air, is diminifhed by refpiration; but the diminution proceeds to a much greater degree, for almoft the whole quantity of elaftic fluid will be reduced to a fmall proportion of carbonic acid gas; and if the experiment be performed on limewater, the whole quantity of oxygen air will difappear. By repeatedly performing the experiment in this manner, it has been found that a healthy middle-aged man will entirely confume two gallons of pure oxygen air in about five minutes time *. But in this cafe the oxygen air is confumed fafter than is neceffary for the ufual fupport of life; and, in fact, if the fame quantity of it be mixed with an equal quantity of azotic

* Amongft the various ways of producing oxygen air, it frequently happens, as we have already hinted, that acid vapours, or other volatile fubftances, are mixed with it; and in that cafe the animal which is confined in it may feel an oppreffion on his lungs, or he may even be fuffocated, when, by the teft of nitrous gas, that air will actually appear to be much better than common air.

gas,

gas, it will then laſt as long again, *viz.* about ten minutes. It is therefore evident, that as the azotic gas is abſolutely incapable of aſſiſting reſpiration, the mixing of it with the oxygen air produces no other effect than that of preſenting a ſmaller quantity of oxygen to the ſurface of the lungs in each inſpiration. It is for the ſame reaſon that oxygen air is conſumed faſter, and that common air is vitiated ſooner, when reſpired under an increaſed, and ſlower when reſpired under a diminiſhed atmoſpherical preſſure.

THE air which is expelled from the lungs after every inſpiration, whether it be oxygen or atmoſpherical air, contains, beſides the portion of carbonic acid gas, a conſiderable quantity of aqueous vapour, which, in cold weather, is manifeſted by its condenſation as ſoon as it comes out of the mouth; for air can hold in ſolution a much greater quantity of water when hot than when cold.

Factitious Airs.

The breathing of pure oxygen air is generally, if not always, attended with an increase of heat, especially about the lungs, and a quickening of the pulse; but on some individuals those effects are increased to such a degree as to produce fevers, inflammation of the lungs, and even consumptions, whilst with other individuals they are moderate, temporary, and even salutary. But I shall endeavour to impress the reader's mind with a clearer idea of those phenomena, by subjoining a short account of the principal experiments that have been performed relatively to this interesting part of our subject.

Dr. PRIESTLEY is, as far as I know, the first person who had the curiosity of breathing oxygen air. " I have," *says he*, " gra-
" tified that curiosity, by breathing it,
" drawing it through a glass-syphon, and
" by this means I reduced a large jar full
" of it to the standard of common air. The
" feeling of it to my lungs, was not sen-
" sibly different from that of common air,
" but

"but I fancied that my breaſt felt peculiarly light and eaſy for ſome time afterwards *."

The following experiment was performed, with great accuracy, before a philoſo-
phical ſociety of gentlemen, at Dr. Hig-
gin's houſe, in the year 1794.—Nineteen
pints of pure oxygen gas were put into a
receiver which ſtood inverted in lime-
water. A tube proceeded from the upper
part of the receiver to the mouth of the
experimenter, a healthy man of about
twenty-two years of age, who, after having accurately ſtopped his noſtrils, and having expired as much air from his lungs as
he poſſibly could in a bent poſture of the
body, took the end of the tube in his
mouth, and began to breathe the oxygen
air in a natural and ſlow manner, during
which the receiver was permitted to play
freely up and down in the lime-water, in
order to prevent any increaſe or decreaſe of

* Experiments on Air, &c: vol. ii. p. 102.

preſſure

pressure on the lungs. An assistant was employed to keep the lime-water in continual agitation, in order to promote the absorption of the carbonic acid air that was formed in the course of the experiment. The bulk of oxygen air was visibly diminished at every inspiration, and the lime-water became turbid. The whole of the oxygen air was consumed in six minutes time, and the experimenter stopped only when the lime-water came to his mouth. " During the respiration his pulse (which,
" previous to the experiment, was only
" sixty-four) quickened to ninety beats in a
" minute, and was considerably increased
" in fulness and strength; but he felt no
" inconvenience whatever.

" THE vessel being immediately charged
" again with nineteen pints of gas, he re-
" spired these also, and consumed them en-
" tirely in six minutes. His pulse was in-
" creased to 120 beats in a minute, and
" was vigorous withal. He felt no in-
" convenience, but had a sense of unusual
" warmth

" warmth in his lungs. In one hour after
" the experiment his pulfe returned to
" fixty-four *."

Dr. Beddoes found the breathing of oxygen air extremely hurtful. " To my own " lungs," *fays he*, " it feels like ardent fpi-
" rit applied to the palate; and I have
" often thought I could not furvive the in-
" fpiration of oxygen air, as it is driven
" from manganefe by heat, many mi-
" nutes †."

A single infpiration of oxygen air may be kept in the lungs much longer than an infpiration of common air.

When oxygen air is mixed with common air, and is then breathed in that diluted ftate, the lungs are lefs affected with

---

* Minutes of the Society for Philofophical Experiments and Converfations, page 146.

† Confiderations on the Medicinal Ufe of Factitious Airs, vol. i. p. 14.

the sensation of heat, nor is the pulse quickened so much as when pure oxygen is used; yet in this diluted state the oxygen air has been found beneficial in a variety of cases, which will be mentioned in the sequel. We shall likewise mention the proportion of the two elastic fluids, which has been found to answer best for each particular case; but in the present chapter it will be necessary to state the effect which the breathing, or the action, of oxygen air has been observed to have upon particular parts of the animal body, whence proper conjectures may be formed of its general use in the animal economy, and of its application for the cure or alleviation of particular disorders.

That oxygen air is a powerful stimulus to the lungs, has been evinced by various experiments, but by none better than the following, which has been repeated with equal success by different persons:—Some young rabbits were kept under water till every appearance of life, and even a hope of recovery, had vaniſhed; they were then withdrawn, and oxygen air was forced through the mouths of ſome of them into their lungs, whilſt a ſimilar operation with common air was performed on the others: the latter remained dead, whilſt the former recovered. Young dogs and kittens were ſubjected to the like experiment, the general reſult of which was, that the oxygen air brought them to life where common air proved ineffectual. Animals thus apparently deprived of life have frequently revived by only being placed in a veſſel full of oxygen air, without forcing it into their lungs. From this fact we derive a powerful method of reſtoring ſuſpended animation.

RABBITS, dogs, kittens, and birds, have been often confined in veſſels full of oxygen air, and have been ſuffered to remain in that quantity of air for various lengths of time. It has been conſtantly obſerved, that they live longer in that, than in an equal quantity of common air. But whenever the experiment

FACTITIOUS AIRS. 37

experiment has been protracted to a certain length, it has almoſt always been attended with illneſs, with a ſtrong inflammation, and even with death. The diſſection of the animals that have been thus *oxygenated*, has principally exhibited the following phænomena:

THE lungs appear of a florid red colour, often marked on the edges with ſigns of mortification; the heart appears of a florid red colour; the pleura is generally inflamed; the colour of the liver, kidneys, and the blood-veſſels of the meſentery, is more inclining to red than is otherwiſe known to be; their blood coagulates ſooner; their muſcles are more vigorous, and ſhew ſigns of ſtronger irritability.

ANIMALS that have breathed oxygen air, previouſly to their being immerſed in water, will not die ſo ſoon as thoſe which have breathed common air only. The quantity of purer air, which remains in the lungs of the former, is what in great mea

D 3 ſure,

fure, if not entirely, contributes to the prefervation of their lives.

AFTER having defcribed, in the preceding paragraphs, the principal phænomena, which are produced by the refpiration of pure, or nearly pure, oxygen air, it will be hardly neceffary to add, that a mixture of common and oxygen airs, or of azotic and oxygen airs, muft produce phænomena analogous to thofe which have been mentioned above, but nearly proportionate to the quantity of oxygen air which is contained in the mixture. There is, however, a remarkable circumftance, which muft be carefully attended to, as being of the utmoft confequence in the application of oxygen air to medicinal ufes. This circumftance is, that whilft the refpiration of pure oxygen air, or of fuch air as contains a great proportion of oxygen, is attended with inflammation and other bad confequences, the refpiration of common air a little improved by the admixture of a moderate proportion (as for inftance, one-15th,

15th, or even one-20th) of oxygen air, is attended with remarkably falutary effects.

THE inhalation of fuch diluted oxygen air, or we may call it improved atmofpherical air, for about 10 or 15 minutes a day, has been found to produce a florid colour in the face, to conciliate fleep, to ftrengthen the organs of digeftion, to promote circulation, to ftrengthen the pulfe, &c.

HOWEVER ftrange and unaccountable thofe effects may at firft fight appear, efpecially to thofe who are not converfant in philofophical inveftigations, the facts are certainly true, and a fimple reflection may contribute to diffipate the wonder; namely, that people of all defcriptions, but efpecially fuch as are weak and emaciated, derive a confiderable degree of exhilaration and improvement by a fhort excurfion out of a town, or of a houfe, when the fuperior purity of the country air,

above

above that of the town, is not equal to that which is produced by mixing common air with even one-twentieth of its bulk of oxygen air. But we ſhall have occaſion to notice this circumſtance again in the ſequel.

CHAPTER III.

Phænomena ariſing from breathing other aerial Fluids, beſides the Common and the Oxygen Airs.

IT has already been noticed, that of the various ſorts of elaſtic fluids, two only, *viz.* the common and the oxygen airs, are capable of aſſiſting reſpiration, from which it may be naturally deduced, that by the admixture of any other gas, either of thoſe two will be rendered leſs reſpirable in different degrees. But this diminiſhed goodneſs of the reſpirable airs, this mixture of
 - reſpirable

FACTITIOUS AIRS. 41

respirable and unrespirable aerial fluids, has proved beneficial in a variety of medical cafes; hence many experiments have been made for the purpose of ascertaining the mixtures that are more applicable to any particular cafe, and likewife the phænomena which arise from the respiration of those mixed gasses.

I WOULD not be understood to assert or think that the action of the unrespirable gasses consists merely in lowering the quality of common air, or of oxygen air; for that purpose could be more commodiously answered by breathing a certain quantity of common air longer than in the usual way. The fact is, that, besides rendering the common or oxygen air less respirable, each particular gas imparts peculiar and remarkable properties to the mixture, which mixtures are of course applicable to particular cafes. With respect to those mixtures, much has already been ascertained; but a great deal more remains to be examined and tried under a variety of circumstances, to

which

which object we muſt look forward with anxious expectation.

It has been repeatedly aſſerted and denied, that pure and unmixed hydrogen, or inflammable gas, may be reſpired with impunity for a conſiderable time, and many experiments are related to prove each of thoſe contradictory aſſertions. The equivocal reſults of thoſe experiments ariſe from two cauſes, *viz.* from the variable nature of the gas, and from the different quantity of common air, which remains in the lungs, mouth, &c. of the animals that are ſubjected to ſuch experiments.

Inflammable gas, in the common way of producing it, is ſeldom very pure; but even when that is the caſe, its coming into contact with the lungs is naturally prevented by the common air, which remains in that organ previouſly to the application of the inflammable gas, the latter being much lighter than the former. By a ſtrong expiration in a bent poſture of the body, the common

common air may, in great meafure, be expelled; but even in that cafe a certain quantity of it unavoidably remains in the mouth, wind-pipe, &c.

OF the different forts of inflammable gas, that which is obtained by paffing the fteam of water over red hot iron feems to be the leaft offenfive. Next to this is the gas which is obtained from iron and diluted vitriolic acid. The other fpecies are more variable in their quality; but they are all incapable of affifting refpiration; and if a perfon will carefully expel as much air from his lungs as he poffibly can by a forced expiration in a bent pofture, and will then apply his mouth to a veffel, or to a tube that communicates with a veffel, full of pure inflammable gas, keeping his noftrils ftopped at the fame time, he will find, after about three or four infpirations, that the florid colour of his face is vanifhed, and his ftrength is fo far diminifhed as to prevent the profecution of the experiment. Having myfelf been more than once witnefs ness to this experiment, I have always observed an evident change of colour in the face of the experimenter after the second inspiration.—The gas had been extracted from iron and diluted vitriolic acid.

INFLAMMABLE gas may be rendered less noxious by agitation in water.

WHEN this gas (meaning that which is obtained from the vapour of water and red hot iron, or from iron and diluted vitriolic acid) is mixed with about an equal quantity, or even a smaller proportion of common air, it may then be breathed with safety for a considerable time; and it is remarkable, that the lungs are affected by it with a peculiar sensation of levity. This singular property has rendered it useful and beneficial in inflammations of the lungs, convulsive coughs, &c. where the object is to diminish the irritability of the parts affected. During this operation the face will be found to grow dark or livid, but the natural colour will be speedily recovered by afterwards

afterwards breathing the common air in the usual way.

The hydrocarbonate, *viz.* that species of inflammable gas which is produced by passing the steam of water over the surface of red hot charcoal, is much more pernicious to the lungs. Animals will die much sooner in this than in the above-mentioned species of inflammable gas. Sometimes two or three inspirations of pure hydrocarbonate are sufficient to occasion the death of the animal.

The active quality of this gas is perceivable even when diluted with 20 or 30 times its own bulk of common air. A person who breathes it in that diluted state for about a quarter of an hour, is generally made sick and vertiginous; feeling at the same time a sensation of cold throughout his whole body; his lips become blue, the face livid, and the pulse feeble, though frequent; but the sensibility of the lungs is considerably diminished by it, on which account

account it has been adminiſtered in various caſes with advantage to the patient. Some patients, after the reſpiration of this diluted gas, have experienced ſuch levity or inſenſibility about the region of the lungs, as to remain for a time entirely free from pain.

It is remarkable, that the ſickneſs, dizzineſs, or, in ſhort, the bad effects of the diluted hydrocarbonate, frequently come on after the operation, and ſometimes come on and go off two or three times repeatedly, at the interval of an hour or longer; which ſhews that this ſort of gas can hardly be adminiſtered with too much care and caution.

Pure carbonic acid gas is likewiſe very pernicious to the lungs. Sometimes one or two inſpirations of it have been quite ſufficient to kill an animal; and, indeed, animals will die in carbonic acid gas, and likewiſe in hydrocarbonate, much ſooner than if they did not reſpire at all, or if they were

were immerfed in water, which proves that fome noxious principle is introduced by thofe gaffes into the body.

Of the animals, thofe which have large lungs in proportion to their bulk, and are formed to live in the air, are fooner affected by this gas; thus the birds have in general been found to die fooneft in carbonic acid air; the dogs come next, then the cats, then the amphibious animals, and laftly, the infects *. If they are not left too long in this gas, they will, in general, revive, by being removed into the common air. When they die in it they fhew no ftruggles. By being frequently expofed to this gas, the animals may be fo habituated as not to be killed by it fo foon as others that were never expofed to it.

The following are the appearances which have been more commonly obferved on the diffected bodies of the animals that have been killed by carbonic acid gas.—The lungs are a little collapfed, fhewing a few

* Bergman *de Acido Aereo*, fect. 26.

inflamed

inflamed places. The right ventricle and right auricle of the heart, the pulmonary artery, the *vena cava*, the jugulars, and the veffels of the brain, are turgid with blood; but the pulmonary veins, the *aorta*, the left ventricle, and left auricle of the heart, are moftly flaccid. The mufcular fibres of the body are found deprived of irritability, fo that even the heart, extracted whilft the body is ftill warm, fhews no figns of irritability *.

FISHES die in a few minutes time, in water impregnated with carbonic acid gas †.

WHEN this gas is diluted with twice or three times its own bulk of common air, it may then be breathed for a certain time, but not nearly fo long as the mild forts of inflammable air fimilarly diluted.

PURE azotic gas is about as deleterious as the inflammable gas from iron and di-

---

* Bergman *de Acido Aereo*, fect. 26.
† Prieftley's Exp. and Obferv. vol. ii. fect. 13. N° 3.

luted

luted vitriolic acid; yet the animals that are confined in it until they appear to be dead, will, on being withdrawn, recover more frequently than thofe which are confined in the inflammable gas.

The artificial gaffes have likewife been breathed in combinations of three or four at a time, one of them always being either the common or the oxygen air; but it does not appear that thofe triple or quadruple mixtures have been tried in a great variety of cafes.

In the refpiring of combined gaffes, due regard muft be had to their fpecific gravities, as this circumftance is often the caufe of phænomena that are erroneoufly attributed to other fources. The difference between the fpecific gravities of the common, the oxygen, and the azotic, airs, is indeed trifling; but the inflammable and the carbonic acid gaffes differ confiderably from the reft, and efpecially from each other; the former being a great deal lighter,

lighter, and the latter much heavier, than common air. If the inflammable, the carbonic acid, and the common or the oxygen, airs, be not well mixed together in a veffel, they will remain feparate for a confiderable time in their refpective places, *viz.* the carbonic acid air in the loweft part, the common in the middle, and the inflammable in the higheft part of the veffel; but even when they are well mixed together, they always fhew a tendency to feparate, fo that after a fhort interval each of them will be found lefs mixed in its refpective place.

IT is hardly neceffary to add, that the fame peculiarity of fituation muft alfo take place within the lungs, and that this is, perhaps, the fole caufe which renders the carbonic acid gas more noxious than the inflammable gas, and the heavy fort of inflammable gas, *hydrocarbonate*, more offenfive than the lighter fpecies of it.

CHAPTER IV.

Phænomena arising from the Application of the abovementioned elastic Fluids to other Parts of the Animal Body besides the Lungs.

IT has been found that the pores of the skin imbibe and expel a small quantity of air, and it is said, that in equal times they will absorb a much greater quantity of oxygen, than of common, or of any other, air.

DIFFERENT sorts of elastic fluid were separately injected into the cellular membrane of animals, through incisions made in the skin, and the apertures were closed immediately after. The appearances, as observed by Dr. Maxwell *, and confirmed by others, were in general as follows:

* See his Thesis, Edinburgh, 1787.

Common air swelled or puffed the animal, rendered it uneasy for a day or two, after which the swelling began to decrease, and vanished entirely at the end of about three weeks.

Oxygen air swelled the animal, and rendered it somewhat uneasy for a short time; the uneasiness, however, soon vanished, the animal became unusually lively, and the swelling disappeared much sooner than in the case where common air had been used.

Azotic gas swelled the animal, and rendered it dull, by superinducing a sort of stupor, which, in a few days time, degenerated into convulsions, and at last killed the animal.

Carbonic acid gas was rapidly absorbed, and seldom produced any slight and temporary uneasiness.

Hydrogen gas swelled the animal, produced heaviness and shiverings; but the swelling

swelling disappeared sooner than in the case of common air.

Mr. GIRTANTER is said to have injected azotic gas into the jugular vein of a dog, in consequence of which the animal died at the end of twenty seconds. On opening its thorax, the pericardium, and the heart, the right auricle and right ventricle were filled with black blood; the left ventricle was of its ordinary dark colour; the heart and muscles had lost their irritability almost entirely. A similar experiment being made with carbonic acid gas, instead of azotic gas, nearly the same phænomena took place.

BLOOD recently taken from the veins of an animal, and exposed to the common air, becomes of a bright red colour; and if exposed to oxygen air, its colour will become still brighter, and the oxygen air will be diminished, and partly converted into carbonic acid air. On the contrary, if the blood thus brightened, or the blood taken from the arteries of an animal, which is well known

known to be of a florid red, be expofed to any of the unrefpirable gaffes, its colour will be darkened prefently, and a fmall part of the elaftic fluid will be abforbed. It is to be remarked, that thofe effects take place even when an animal membrane, as a piece of bladder, intervenes between the blood and the refpirable or unrefpirable elaftic fluids *. Even the colour of the flefhy parts is made to incline more towards a florid red by the action of oxygen air.

That the oxygen air acts as a ftimulus on other parts of the body, as well as on the lungs, is clearly proved by the following often repeated experiment: A blifter being formed on the hand, or a finger, by the application of the ufual plaifter of cantharides, the fkin was cut off, and the hand was immediately introduced into a veffel full of oxygen air: the confequence was, that the experimenter felt a very acute pain. The hand was then removed into a veffel full of

---

* Prieftley's Exp. and Obf. vol. III. fect. 5.

carbonic

carbonic acid gas, the action of which removed the pain in a very short time. On the hand being expofed to the common air, a degree of pain returned, and on being, as at firft, placed in oxygen air, the pain became acute.

The contact of inflammable gas does neither accelerate nor retard the putrefaction of animal matter.

When the ftream of carbonic acid air is iffuing out of a fmall aperture, as that of the tube of the phial in which this gas is ufually produced from calcareous earth and diluted vitriolic acid, if the mouth or noftrils be prefented to it, they will be affected with a peculiar, and rather pleafing, pungency.

This gas is poffeffed of confiderable antifeptic power. And for this property, it is adminiftered to the animal body either internally or externally, and feparate parts of animal or vegetable fubftances may be preferved in it for a confiderable time.

It is applied internally to the ſtomach, or externally, either in the aerial form, or combined with water and other ſubſtances. Many fluid or ſolid bodies derive their antiſeptic property from their containing this gas in conſiderable quantity; ſuch are liquors in a ſtate of vinous fermentation, ripe fruit, certain mineral waters, &c.

Fruit may be preſerved ſeveral days longer in carbonic acid than in common air. This is alſo the caſe with animal fluids, or with pieces of meat that are not very large, but they are apt to looſe their flavour. Large pieces of meat are ſaid to have been preſerved for ſeveral days longer than in the uſual way, by only waſhing them three or four times a day in water ſtrongly impregnated with carbonic acid air.

Distilled water, or water that has been deprived of its air by boiling, will, in forty days time, and in a temperate atmoſphere, abſorb, without needing any agitation,

tion, about $\frac{1}{14}$th of its bulk of oxygen air, whereas of common air it will abforb about the half of that quantity, *viz.* $\frac{1}{28}$th part. It will abforb in a few hours time a quantity of carbonic acid gas little greater than its own bulk; but a cold temperature and an increafed atmofpherical preffure will enable it to abforb a much greater quantity of that gas. Of inflammable gas it will abforb about as much as it does of common air, *viz.* $\frac{1}{14}$th part of its bulk.

This abforption of elaftic fluids by water is much expedited by agitation of the latter in the former.

CHAPTER V.

Theory of the Nature of Aerial Fluids, and of Respiration.

THAT respiration and life can not be maintained without atmospherical air, is a fact known to the philosophers of the remotest antiquity; but their ideas of the use of air in respiration were vague, and unsupported by experiments. On the revival of learning in Europe, and especially after the sixteenth century, the scientific inquiries of philosophers, physicians, and chemists, ascertained that the air was subservient to other natural as well as artificial processes, besides respiration; and likewise that there actually existed various species of air, some of which were highly noxious *. The progress and dissemination of science gradually added new articles to

* See the works of Van Helmont and Dr. Mayow.

the stock of knowledge relative to the aerial fluids; but the great improvements, the surprising discoveries, which have produced a total revolution in this branch of natural philosophy, were reserved for the present age, and are undoubtedly due to the labours of modern philosophers.

It is entertaining to peruse the works of authors previous to the late discoveries, and to observe how near the ideas of some of them approached the modern theory of respiration. Hippocrates considered air as one of the aliments of the body. Dr. Mayow asserts, that some nitre, or aerial spirit of nitre, enters the body through the lungs, and furnishes the animal spirits at the same time that it communicates heat to the blood*.

Dr. White supposed that the stimulating quality of the air is necessary to keep the heart in motion. Mr. Hewson, observ-

* See his work, printed at Oxford in the year 1674, under the title of *Tractatus quinque Medico-Physici*.

ing that the blood has a more florid red appearance in the left, than in the right, auricle of the heart, concludes with faying, that as the change of colour in blood out of the body is occafioned by the contact of air, fo it may be prefumed that the fame change within the body is occafioned by air alfo, and that the change takes place in the lungs.

DR. PRIESTLEY formed a very ingenious hypothefis concerning the ufe of air in refpiration, which he eftablifhed by a train of well-conducted experiments on the then prevailing phlogiftic theory. The principal law of this hypothefis is, that the air ferves to abforb the fuperfluous phlogifton from the blood through the lungs, and that the more or lefs florid rednefs of the blood depends on the different quantities of phlogifton in it *. The phlogifton, however, or principle of inflammability, is not

* For a full explanation of this hypothefis fee the Doctor's Exper. and Obf. vol. III. fect. 5; or the Phil. Tranf. vol. LXVI.; and likewife Dr. Crawford's work on Animal Heat and Inflammation.

a real,

a real, but a suppofed, agent in nature, which, for want of better information, was applied to explain moſt of the phænomena of combuſtion, decompoſition, and (by Dr. Prieſtley's ingenuity) of refpiration. But the prefent ſtate of knowledge being, in confequence of very recent difcoveries, fufficient to account for the abovementioned phænomena in a fimpler, and, of courfe, a more natural way, the fuppoſition of the phlogiſtic principle is become altogether fuperfluous.

OF this new or antiphlogiſtic theory, which may be feen at large in a variety of recent publications, and of the difcoveries which gave rife to it, I ſhall briefly mention ſuch particulars only, as may be of ufe in elucidating the action of the aerial fluids on the human body. As for the facts upon which its feveral parts are eſtabliſhed, and likewife for the objections which have been made to it, I muſt refer the reader to the works of other authors [*].

[*] See Lavoifier's Elements of Chemiſtry, Dr. Prieſtley's pamphlet, entitled, Experiments and Obfervations relating to the Analyſis of Atmofpherical Air, &c.; Fourcroy's Chemiſtry, &c.

THIS

This theory is at prefent almoft univerfally adopted by perfons of the firft rank in philofophy, and daily experience is continually throwing new light upon it; yet it muft be confeffed that it is by no means free from doubts and difficulties. It is in confequence of thofe deficiencies, and on account of the uncertainty, which is infeparable from the nature of hypothefes, that I have carefully feparated the knowledge of facts from the fuppofition of their caufes. The former have been arranged in the preceding four chapters, and any perfon may account for them in the manner he likes beft; but it was deemed neceffary, at the fame time, to add the moft fatisfactory explanation which can be fuggefted by the prefent ftate of knowledge, and this explanation will be found in the prefent chapter.

The fenfation of heat is fuppofed to be produced by a peculiar fluid called *the caloric*, or elementary heat; a fluid extremely fine, penetrating, and fo light that its weight cannot

FACTITIOUS AIRS. 63

not be eftimated. All forts of bodies are expanded by the addition, and contracted by the abftraction of caloric. The acceffion of it to the human body produces the fenfation of heat, and the feparation of it produces the fenfation of cold. Thus when we touch a fubftance which is of a lower temperature, viz. colder than our bodies, that fubftance, by robbing us of a portion of caloric, will excite the fenfation of cold; and on the contrary, if the fubftance be hotter than our bodies, it will excite the fenfation of heat, by adding caloric to our bodies.

WHEN a number of bodies of different temperatures are put together, the fum of their quantities of caloric will be difperfed amongft them in fuch a manner as to render them all of the fame temperature, fo that a thermometer will be found to indicate the fame precife degree of heat in any one of them. But it muft be remarked, that though the temperature be the fame, yet the abovementioned fum of elementary heat will not be divided equally amongft the bodies,

bodies, unless the bodies be of the same sort, as, for instance, three or four parcels of water, or of mercury, &c.; but some bodies will imbibe more and others less of the caloric, in order to be raised to the same temperature, or apparent degree of heat; and this peculiar disposition in any particular body is called its *capacity for containing* caloric. This property of bodies may be rendered more intelligible by an example or two. Suppose that a pint of water, at 100° of heat, be mixed with another pint of water at 200° of heat, the heat of the mixture will be nearly 150°, *viz.* an arithmetical mean between the two temperatures; but if a pint of water at 100° of heat be mixed with a pint of quicksilver at 50° of heat, the heat of the mixture will be found to be 80°, (*viz.* greater than 75°, which is the arithmetical mean) which shews that either the quicksilver or the water, has imbibed more than its equal share of caloric, in order to have its temperature raised to the common degree of sensible heat. On the other hand, if the degrees of heat be

reversed,

reverſed, *viz.* the water at 50° be mixed with an equal bulk of quickſilver at 100°, the temperature of the mixture will be 70°, which plainly ſhews, that water abſorbs more heat than quickſilver; and as the difference between their original temperatures and the temperatures of the mixture in the firſt and laſt caſe is as two to three, we therefore ſay, that the abſolute heat of mercury is to that of an equal bulk of water as two to three; *viz.* "that the compa-
"rative quantities of their *abſolute* heats
"are reciprocally proportionable to the
"changes which are produced in their
"*ſenſible* heats, when they are mixed to-
"gether at different temperatures *."

SIMILAR experiments performed on a variety of bodies ſhew, that unequal quantities of abſolute heat muſt be communicated to them in order to raiſe their temperature, or apparent heat to the ſame degree.

* Dr. Crawford on Animal Heat and Inflammation; in which work a full explanation of the doctrine of heat will be found, together with a table of the comparative heats of different bodies.

It is in confequence of their various capacities, that whenever bodies of different fpecies are brought together, a change of temperature is generally produced. Thus, if you mix fpirit of wine and water, the mixture will become hotter than the ingredients were before. A much greater degree of heat will be produced by mixing water with vitriolic acid; and, on the other hand, if fal ammoniac de diffolved in water, a confiderable degree of cold will be produced.

In moft fubftances a total change in their ftate of exiftence is produced by the fuperaddition of caloric; thus water is gradually changed from its folid ftate of ice, into a fluid, and then into an elaftic fluid, called vapour, by the addition of different degrees of caloric. And it muft be remarked, that this change of ftate in bodies is attended with a change of capacity for containing caloric; the lefs denfe ftate containing the greateft quantity of caloric. Thus water in the fluid ftate contains lefs caloric

caloric than when it is reduced into vapour, and more than when it exifts in the form of ice.

The aerial fluids are fuppofed to be combinations of certain fubftances with caloric. Oxygen air confifts of a fubftance, *fui generis*, which is called oxygen, combined with caloric, and, in all probability, with the matter of light alfo.

Azotic gas confifts of a particular fubftance, called *azote*, and caloric. Common air confifts of azotic gas and oxygen air, in the proportion of 73 parts of the former to 27 of the latter. By a mixture of thofe elaftic fluids in the faid proportion, an aerial fluid is formed exactly like the atmofpherical air *.

Hydrogen gas confifts of a particular fubftance, called *hydrogen*, and caloric. As

* In general the atmofpherical fluid contains a variety of extraneous particles, but they hardly ever exceed the hundredth part of the whole, and feldom amount to that quantity.

for the particular fubftances which are frequently found in the hydrogen gas, fuch as phofphorus, particles of iron, &c.; they muft be confidered as extraneous matters fufpended or diffolved in the gas, but not effential to its conftitution.

CARBONIC acid gas confifts of a peculiar fubftance, called *carbon*, or the conftituent part of charcoal, and oxygen air, in the proportion of feven of the former to eighteen of the latter.

WATER, which has long been efteemed an elementary fubftance, incapable of decompofition, has been found to confift of hydrogen gas and oxygen gas, in the proportion of three of the former to feventeen of the latter. By the combuftion of thofe elaftic fluids water is actually formed; and, on the contrary, water may be reduced into thofe aerial fluids, by placing it, under certain circumftances, in contact with bodies that attract one of its components, or by the action of electricity *.

* See Fourcroy's Chemiftry, the third volume of my Electricity, and the Phil. Tranf. for 1797, P. I.

Factitious Airs.

Combustion confifts in the abforption of the bafe of oxygen air, *viz*. the oxygen, by bodies that are faid to be combuftible, and fetting free both the caloric and the light, which, as has been mentioned above, are the two other components of oxygen air. Agreeably to this definition, we muft confider as combuftions not only the burning of coals and other fuel, as is ufually done in our chimneys, but alfo the calcination of metals, and refpiration itfelf, fince in both thofe proceffes an abforption of oxygen, and an evolution of caloric, take place.

If the calcination of a metal (which is now called *oxygenation* of the metal) is carried on flowly, as by merely expofing certain metals to the atmofphere, then the caloric and the light, which is feparated from the oxygen portion of the atmofphere, is too little to affect our fenfes, and we can only obferve, after a certain time, that by having abforbed a quantity of oxygen, the metallic fubftance has loft its combuftibility

lity *(viz.* its attraction for oxygen) and has aſſumed a different appearance, together with an increaſe of weight. If the oxygenation be carried on in a quick manner, as when an iron wire is made very hot in oxygen air, then both the caloric and the light become manifeſt.

WHEN the metal has abſorbed as much oxygen as its nature admits of, it is then ſaid to be incombuſtible, or completely oxydated. But if by any means the oxygen be ſeparated from it, then the metallic oxyde will be converted again into a metallic ſubſtance fuſceptible of combuſtion.

IN the combuſtion of animal and vegetable ſubſtances, which confiſt of various component articles, the proceſs is accompanied with peculiar phænomena, which vary with the nature of the combuſtible, the quicknefs of the combuſtion, and other circumſtances. Thus in the burning of wood, the oxygen of the atmoſphere is abſorbed, the caloric and the light are difengaged,

gaged, the carbon of the wood combines with a portion of the oxygen, and forms carbonic acid gas, the evolved caloric converts the aqueous part of the wood into steam, and so forth.

A VARIETY of phænomena may be obviously explained upon the basis of this doctrine.

WE may easily comprehend why no sort of combustion can take place where no oxygen air exists; as also why every sort of combustion will proceed rapidly in pure oxygen air, and much less so in common air; for the latter contains only a small proportion (*viz.* about one quarter) of oxygen air. Thus may other processes be easily reconciled to, or explained by, this theory. But it is now time to examine the phænomena of respiration.

THE uses of respiration are various and important. They may be divided into mechanical and chymical. Of the mechanical, such as the voice, the cough, &c. no notice will be taken in this work. The other uses principally consist in furnishing the body with a constant supply of oxygen, and probably in exonerating the blood of the superfluous carbon and hydrogen.

In the process of respiration a decomposition of the air takes place in the lungs. The blood, in its passage through that organ, absorbs the oxygen of the common air, disengages the caloric, and leaves the azotic gas, with a small residuum of oxygen air. The blood, therefore, does not imbibe the oxygen air, but the oxygen alone, *viz.* the basis of oxygen air, divested of that quantity of caloric which was necessary to give it the aerial form. The caloric which is set free in this process, by dispersing itself through the body, keeps up its temperature, and forms the origin of animal heat. However, this part of the theory which relates to the formation of animal heat, is embarrassed with difficulties, which will be noticed presently.

THE

THE carbonic acid gas, which is formed in the procefs of refpiration, is fuppofed to derive its origin from a quantity of carbon, which, being difcharged from the blood, combines with a portion of the oxygen air.

THE watery vapour which is expelled with the air that is expired from the lungs, is fuppofed to be formed in that organ by a combination of oxygen with a quantity of hydrogen, which is likewife difcharged from the blood. But it is not unlikely that both the carbonic acid gas and the water, inftead of being formed in the lungs, may come out of the blood, through the exhaling pores of that organ, ready formed; the blood having originally received it in that ftate from the chyle, &c.

THE air then which is expired from the lungs, contains a fmaller quantity of oxygen air than it did before, but it contains alfo fome carbonic acid gas, and fome water,

water, in the form of vapour*.—Let us now examine the different parts of this theory.

It is evident, from the foregoing facts and explanations, that the oxygen air is the only fluid capable of affifting refpiration and combuftion, and that it is indifpenfably neceffary to animal life, fince the common air is ufeful only on account of the oxygen it contains.

* There are fome modern philofophers, who explain the phænomena of refpiration without admitting the abforption of oxygen by the blood. The blood, they fay, in paffing through the lungs, acquires a vermilion red colour, becaufe it depofits a portion of its carbonated hydrogen upon the air; and it becomes again dark in the courfe of circulation, becaufe it combines with a frefh quantity of carbonated hydrogen. At the fame time the oxygen of the common air which enters the lungs, by combining with the carbon and with the hydrogen, forms the carbonic acid gas with the former, and the watery vapour with the latter.—It may be eafily perceived, that by only changing the name of carbon into that of phlogifton, this explanation may, in a great meafure, be made to coincide with Dr. Prieftley's hypothefis.

THE

FACTITIOUS AIRS. 75

THE mixture of nearly one part of oxygen air and three parts of azotic gas, which forms the atmofpherical fluid, is, in all probability, the beft proportion of ingredients for the maintenance of life ; fince we find that with a fmaller proportion of oxygen, not only the refpiration becomes unpleafant and laborious, but debility, convulfions, and other bad effects are produced; and on the other hand, that bad fymptoms of another fort are brought on by a greater proportion of it, fuch as a preternatural heat, feverifh pulfation, pains, inflammations, &c.

THE phænomena of refpiration and of combuftion are not only analogous, but they illuftrate each other in an admirable manner. In atmofpherical air a candle gives light fufficient for ordinary purpofes. In a lefs pure atmofphere the light becomes too dim ; and in pure, or nearly pure, oxygen air, the candle will indeed give a much brighter light; but it will wafte fo very faft, as not to laft perhaps the twentieth part of the time it will in common air.

THAT

That the blood abforbs the oxygen of the atmofpherical air in the act of refpiration, is a propofition which a variety of experiments and analogies feem to prove beyond all doubt. When blood, recently taken out of the veins of an animal, is enclofed in a piece of bladder, and is thus expofed to common air, or to oxygen air, it acquires a florid red colour, and part of the oxygen air is abforbed. The fame thing takes place within the body, *viz.* the air which is expired contains a fmaller proportion of oxygen than it did before, and the blood which returns from the lungs to the heart, and thence proceeds through the arteries, is found to have acquired a bright rednefs in its paffage through the lungs; it is therefore natural to conclude, that the blood has abforbed the oxygen through the pores of the thin membrane, which feparates it from the air in the cells of the lungs *.

* This membrane is certainly much thinner than common bladder. Dr. Hales conjectured the thicknefs of the former to be the thoufandth part of an inch.

THE probability of this conclusion is corroborated by strong collateral proofs; as by observing that the arterial blood of animals that have been suffocated, or that have died for want of oxygen air, is far from being of its usual florid red colour; as also by observing, that when a quantity of blood is confined in a vessel full of air, the air is not so quickly contaminated or deprived of its oxygen by the presence of arterial, as by that of venous blood. And it is even asserted, that a quantity of blood taken out of the carotid artery of a sheep, being confined in a vessel full of azotic gas, improved the gas so as to render it, in some measure, fit for respiration, so that some oxygen must have been imparted to it by the blood *. This experiment deserves to be repeated with particular care.

THE decomposition of air, and the absorption of its oxygen in combustion and oxygenation of metallic bodies, are also analogous

* Medical Extracts, vol. I. p. 70.

logous to the phænomena of refpiration, and confirm the abforption of oxygen by the blood in that procefs.

It is true that a fmall quantity of carbonic acid gas is found in the air in which blood has been confined; but the formation of this gas does not entirely account for the diminution of the oxygen. Befides, it is not improbable, as we faid above, that the carbonic acid gas comes out of the blood ready formed, at the fame time that the blood abforbs the oxygen.

By examining the courfe and ftate of the blood, we find that it preferves its brilliant rednefs through all the channels which convey it from the lungs through the heart, and to the extremities of the body. But in the other veffels, which receive it at the extremities of the former, and convey it through the heart as far as the lungs, the blood is of a dark purple colour. The former courfe is performed through the pulmonary veins, the left auricle and left ventricle

ventricle of the heart, the aorta and its branches. The latter is performed through the branches and trunks of the afcending and defcending cava, the right auricle and right ventricle of the heart, and laftly, through the pulmonary arteries, which convey it to the fpungy cells of the lungs, where its colour is changed, &c.

THE blood, therefore, having acquired the oxygen in the lungs, conveys it as far as the extremities of the branches of the aorta, where the oxygen is depofited, and the blood returns without it through the veins.

IT is difficult to fay under what form is the oxygen combined with the blood, and what becomes of it at the extremities of the arteries where it is left by the blood. For want of direct experimental information concerning this interefting point, we have only the light of analogy and conjecture to lead us in the inveftigation of truth.

IN

IN the combinations of the base of oxygen air with different bodies, such as take place in combustions of every sort and degree, three different effects must be particularly remarked. The first is, that the oxygenation is generally accompanied with colours of different intensity; the red being produced more frequently than any other colour, as is the case with *mercurius calcinatus per se, red lead, crocus martis,* &c. The second is, that by the accession of oxygen a body is always rendered firmer or more compact. Thus, by the combustion of hydrogene and oxygen, water is produced, which is a much heavier and more compact substance than either of its two components; thus also by oxygenation oils are thickened, and metallic bodies are converted into a substance powdery indeed, but whose particles are firmer and harder than the same bodies in their metallic state*. The third is, that a body loses, in great

---

* It is in consequence of the superior hardness of its particles, that *crocus martis* (which is oxygenated iron) and

great meafure, its capacity for containing caloric, and of courfe gives out heat whenever it paffes from a rare into a more compact ftate of exiftence, and *vice verfa*. Thus water contains a great deal more of caloric than ice, but much lefs than fteam; hence when fteam is converted into water, it depofits part of its caloric, *viz.* it communicates fenfible heat to the furrounding bodies, &c.

By an eafy application of thofe facts to the phænomena of refpiration, we are led to conclude, firft, that the rednefs which the blood acquires in the lungs, indicates a real oxygenation of that fluid; fecondly, that the oxygen is flightly attached to the blood, for the blood eafily parts with it at the extremities of the arteries; thirdly, that the oxygen, which is depofited by the blood at the extremities of the arteries, enters into combination with, and gives firm-

and the oxyde of tin (commonly called *putty*) are employed for polifhing the hardeft fteel, glafs, and even agates.

G   nefs

nefs and folidity to, thofe particles of matter which give increment and ftability to the animal frame; fourthly, and laftly, that as the bond of union between the blood and the oxygen is not very ftrong, and as the union of the oxygen with other fubftances at the extremities of the body is much ftronger, therefore it feems evident that the caloric of the oxygen air is not entirely evolved from it in the lungs; but that the greater portion of caloric is evolved at the extremities of the arteries, where the oxygen is more powerfully attracted by other fubftances than it is by the blood in the lungs. Hence it follows, that the origin of animal heat does not exift in the lungs only, but that it takes place, more or lefs, in every part of the body. And this fhews why the whole body is nearly of the fame temperature; whereas, if the caloric were evolved in the lungs only, that part of the body would be much warmer than any other, which is not the cafe *.

WHAT

* I am happy to find that this explanation coincides with the opinion of a very diftinguifhed and recent anatomical

WHAT difpofes the blood to abforb the oxygen in the lungs, and what forces it to depofit that principle at the extremities of the arteries, are queftions which the prefent ftate of knowledge does not enable us to anfwer fatisfactorily. It has been fuppofed that the oxygen is attracted by the ferruginous particles of the blood, and that the rednefs of the blood is to be attributed to the red colour of the oxyde of iron. But fince

tomical writer, who expreffes himfelf in the following words:

" But in reflecting upon this moft difficult of all fubjects, the generation of heat in the living body, many things are to be taken in the calculation, which feem, on the flighteft glance, to be far more important than this depofition of oxygene from the blood. It is a law of nature, to which, as far as we know, no exception is found, that a body, while it paffes from an aerial to a fluid, or from a fluid to a folid form, gives out heat. Now, what is the whole bufinefs of the living fyftem but a continual affimilation of new parts, making them continually pafs from a fluid into a folid form? The whole nourifhment of the body goes on in the extreme veffels, and is a continual affumption of new parts. The extreme veffels are continually employed in forming fome acids, which appear naked in the fecretions; in forming oxyds, since it has been proved by a variety of experiments, that the oxygen is attracted by, and combines with, a variety of other substances independant of iron or metals, I do not see the necessity of attributing the attraction of the oxygen to the ferruginous, more than to other, ingredients of the blood. Nor do I see the absolute necessity of attributing the red colour to the particles of iron, since other substances, in which iron is not concerned, such as the oxyde of mer-

oxyds, as the fat and the jellies of the membranous and white parts; in the various depositions of muscle, bone, tendon, &c. for these are all continually absorbed, thrown off by the urine, and incessantly renewed. They are continually employed in filling all the interstices of the body with a bland fluid or halitus; they are continually employed in forming secretions of various kinds. In performing all this the power of the vessels may do much; but the ultimate effect in each procefs muft be a chemical change, and perpetual changes will produce a constant heat. Place the organ and focus of this animal heat in the centre of the body, and you are embarrassed in a thousand difficulties; allow this heat to arise in each part according to its degree of action, and each part provides for itself." Bell's Anatomy, vol. II. p. 125.

cury,

cury, red lead, &c. owe their redneſs merely to the oxygen which they have imbibed.

It is difficult to account for the formation of the carbonic acid gas, and of the watery vapour in the lungs; for if thoſe fluids be really formed in that organ by the combination of the carbone, and of the hydrogen, with the oxygene of the inſpired air; the whole, or nearly the whole, of the oxygen air would be ſo expended, and little or none of it would remain to be imbibed by the blood. The caloric likewiſe would be employed in the formation of thoſe fluids, inſtead of being difperſed through the body. Is it not therefore more natural and more ſatisfactory to ſuppoſe, that both the carbonic acid gas, and the water, are ſeparated from the blood in the lungs, but not formed in that organ? It is certain that carbonic acid gas is introduced into the ſtomach by the aliments; and it is certain that the chyle conveys it to the blood, why then ſhould we ſuppoſe that there is another formation of this gas in the lungs? As for the

watery

watery vapour, we may account for it in the same manner; and indeed the exudation of water through the internal membranes of the human body, is so generally practised by nature for the purpose of keeping those membranes, &c. soft and pliant, that it would be irregular not to admit the same exudation of water in the lungs also.

The expulsion of putrid effluvia from the body is considered as another office of respiration. This is shewn by the offensive smell of the breath of certain persons, who have no bad teeth to account for it. But it is difficult to ascertain in what cases this may take place, and how far it may extend.

It is with the appearance of probability supposed that the oxygen, which the blood deposits on the various parts of the body, is partly expended in the exercise of muscular motion; since we find, that after unusual exertions of the body, a man breathes faster, and likewise takes in much more air at a time, as if nature endeavoured by that

means

means to recruit what has been lately expended in greater quantity than usual.

The azotic gas, which is the greatest ingredient of common air, is considered as only a diluent of the oxygen air, and as being otherwise passive in the process of respiration. Yet this diluent answers a variety of purposes; the principal of which is, that it exposes a proper quantity of oxygen air to a great quantity of blood, which could not have been the case if the atmospherical fluid had consisted entirely of oxygen. This object is accomplished by the very extensive surface which the lungs present to the air in its numerous cells; for the more numerous the cavities are, the greater is the surface; and, in fact, we find that in those animals that are not much in want of air, and that must frequently suspend their respiration for a considerable time, such as the sea-turtle and the frog, the lungs consist of very few and very large cells.

The great proportion of azotic gas in common air, does alſo adapt that fluid to the purpoſes of vegetation, and other natural proceſſes, the enumeration of which is incompatible with the limits of this Eſſay.

CHAPTER VI.

A general Idea of the Application of aerial Fluids for the Cure of Diſorders incident to the human Body.

CONSTANT obſervation has informed mankind, from time immemorial, that the air of certain places is more or leſs ſalubrious than that of other places; and that the various qualities of the air in different ſituations, are peculiarly favourable to certain conſtitutions. Phyſicians, availing themſelves of this natural variety, have
long

long been in the habit of fending their patients to fuch places as experience and analogy indicated to be more favourable to their refpiration. The fharp air of one place was reckoned good for one diforder, the damp air of a fecond place was efteemed ufeful in other cafes, the pure air of a third was recommended in particular difeafes, and fo on. Howfoever defective and erroneous their knowledge of the real conftitution of the atmofphere may have been, howfoever they may have abufed the application, yet certain it is, that the variety of effects, fuitably to the different qualities of the atmofpherical fluid in different fituations, is attefted by innumerable facts and univerfal obfervation. Previoufly to the late difcoveries, the ideas of phyficians refpecting the different qualities and effects of the atmofpherical fluid, were vague, and generally erroneous. Experience, which fhewed them the advantages that had been obtained in a number of fimilar cafes, was their beft guide, and all befides was doubt and obfcurity. The prefent ftate of knowledge has, in great meafure, measure, dissipated the clouds; since it has not only shewn the reasons upon which certain qualities of the air depend, but has likewise furnished us with the means of procuring airs of opposite qualities, and of any degree of purity, at all times and places, as also of applying them in all the extensive variety of quality, degree of purity, and length of time.

The apparatus necessary for producing the various factitious airs, may be easily derived from the particulars that have been mentioned towards the beginning of this book; but for a general apparatus, that admits of compactness, cheapness, and a sufficiently extensive application, I cannot recommend any better than, or nearly so good as, that which was contrived by Mr. James Watt, engineer, of Birmingham; by means of which the artificial airs, of sufficient purity, may be produced at a very moderate expense, and easier than by any other general method. Those apparatuses are now made for sale, and a printed description,
with

FACTITIOUS AIRS. 91

with neceſſary practical directions for the uſe of its various parts, is given with each apparatus, which ſuperſedes the neceſſity of adding the ſame to the preſent work. I ſhall, neverthelefs, reſerve, for the end of the book, a liſt of the principal precautions which ſhould be attended to in the management of Mr. Watt's, or of any other apparatus of this ſort, to which the practitioner may recur for extempore information.

The artificial elaſtic fluids are applied to the lungs by the way of reſpiration, to the ſtomach and inteſtines, by means of injections, or in combination with fluids, and to the external parts of the body, merely by contact.

Various apparatuſes have been uſed for the reſpiration of factitious airs. The leaſt exceptionable air-holder for this purpoſe, conſiſts of a large glaſs receiver filled with the required ſort of elaſtic fluid, inverted, and ſwimming in water; out of which the air

air is refpired by means of a bent glafs tube, which, paffing with its bent part through the water, projects one aperture above the water within the receiver, whilft its other extremity is applied to the mouth of the experimenter.

INSTEAD of the above-mentioned bent tube, the receiver may have an aperture at its upper end, to which a tube is adapted, air tight, in an horizontal direction. But as this apparatus requires a large tub full of water for the receiver to fwim in, which renders it rather cumberfome, and not very portable, therefore other contrivances have been fubftituted to the receivers. The machine more in ufe for this purpofe, confifts of an oil-filk bag, furnifhed with a fhort wooden tube or faucet, which, when the bag is full of the required aerial fluid, is applied to the mouth of the patient. Thofe bags are filled with the proper aerial fluid by means of a glafs receiver, which, befides its large aperture, has a fmall aperture with a ftopple at the oppofite end. The receiver being

being filled with the required air, and inverted in water, the ftopple is removed from its fmall aperture, and the wooden tube of the bag is applied quickly to it; then by preffing the receivers down into the water of the tub, the air will be forced into the bag. But with the air-holders, which form part of Mr. Watt's apparatus, the operation is rather eafier, for which, fee the defcription of the faid apparatus. The principal imperfection of thefe bags confifts in the fmell of the oil-filk, which proves naufeous, and almoft intolerable to delicate perfons; yet this fmell may, in fome meafure, be removed *.

INSTEAD

* For this purpofe Mr. Watt gives the following directions:—" To free oiled filk from its difagreeable
" fmell, cut it into pieces of the fize wanted for the
" bags, and provide a fmooth table fomewhat larger
" than the pieces of filk, and a flat board of the fame
" fize as the table. Take charcoal frefh burnt in an
" open fire, until it is free from fmoke, extinguifh it by
" fhutting it up in a clean clofe veffel, and reduce it to
" powder. Sift this powder over the table to the
" thicknefs of a quarter of an inch, or more, fpread a
" piece

INSTEAD of oil-silk bags large bladders may be used; but as a bladder is not capable of holding a quantity of elastic fluid large enough for medicinal use, several of them should be had in readiness, each fur-

" piece of your silk upon it, and sift upon that again ano-
" ther layer of your charcoal dust, and thus proceed al-
" ternating the layers of silk, and charcoal, until the
" whole of your silk is deposited; then lay your move-
" able board upon the top of all, and leave the whole
" undisturbed for four or five days. If, upon remov-
" ing the charcoal dust, the silk has not lost its smell en-
" tirely, repeat the process. The charcoal dust is to be
" swept off the silk, and the silk to be washed upon a
" table with a wet sponge until it is clean. The bags
" must then be carefully sowed up, and the seams
" anointed with japanner's gold size, taking care to use
" that kind which does not become brittle when dry.
" Green oiled silk should be avoided, as it is stained by
" means of verdigris, which rots it; the yellowish silk
" is the best.

" It is necessary to observe here, that although oiled
" silk be the best substance known for making the bags
" of, it is very imperfectly air-tight; and although
" charcoal-dust deprives it of smell for the time, yet as
" it can only attract the odoriferous particles from the
" surface, it re-acquires some smell by keeping, but by
" no means equal to what it had at first."

nished

nifhed with a wooden or glafs tube, like the oil-filk bags, through which they may be filled, &c.; fo that when the air of one bladder is exhaufted, a fecond bladder may be fubftituted, and fo on. Several bladders might be eafily made to communicate with each other, fo that through one tube or faucet they might be filled all at once: four or five large bladders thus joined together, would contain about as much air as an ordinary oil-filk bag, which is a quantity, in moft cafes, fufficient for one application, and it would laft about fix minutes. The bladders have likewife an unpleafant fmell, which may alfo, in great meafure, be removed *.

WHETHER the glafs receiver, or the oil-filk bags, or the bladders be ufed, the patient muft always take care to keep his noftrils accurately ftopped whilft he draws

---

* For this purpofe turn the bladder infide out, wafh it well with a weak folution of falt of tartar, then wafh it feveral times over with fair water, fo as to remove every particle of alkali; laftly, wafh it with fpirit of wine.

the air into his lungs at every infpiration, and to open them immediately after, fo as to expel the air from his lungs through the noftrils into the atmofphere at every expiration, and not to return it into the bag or receiver. This operation is not eafily performed by moft perfons, and fome there are who cannot perform it at all; in which cafe they breathe the fame air backwards and forwards to and from the bag. But by this means the air of the bag, even when limewater is contained in it, is contaminated fo quickly as to do more harm than good.

THIS inconvenience, however, is completely obviated by the ufe of a little machine, which is to be interpofed between the mouth of the patient and the faucet of the bag. It confifts of a fmall box of wood, having three apertures, to the two oppofite of which two fhort tubes are faftened; to the third, which is a lateral one, there is an external valve which will only permit the air's going out of the box into the atmofphere. One of thofe tubes is applied to the

the faucet of the bag, and it contains a valve which prevents the return of the air into the bag; the other tube is applied to the mouth of the patient, who has nothing more to do than to hold his noftrils conftantly ftopped, and to breathe in a natural way as long as there is any air in the bag or receiver *; it being eafy to underftand that whenever he infpires, the air will pafs from the bag into his lungs; but that at every expiration, the air will be forced through the lateral valve of the machine into the ambient air.

Of the various forts of elaftic fluids, the carbonic acid gas is the only one that has been fuccefsfully applied to the ftomach or inteftines, and for this purpofe it may be adminiftered two ways, *viz.* either in the aerial form in clyfters, or combined with different fluids and given through the

* There are feveral perfons who, with very little attention, can breathe through the mouth only; when this is the cafe, the keeping of the noftrils ftopped is fuperfluous.

H mouth.

mouth. For the former of those purposes the gas must be first introduced into a bladder by the method already described. For the latter, the gas is either naturally contained in liquors, as in newly fermented liquors, yeast, certain ripe fruits, and mineral waters; or is to be first combined with the required liquors, in which case water is the fluid which is more generally used. This impregnation of water and other liquors with carbonic acid air, may be accomplished by various methods, such as by pouring the liquor backwards and forwards from one vessel to another, over the surface of vegetable substances that are in a strong state of fermentation; or by filling a vessel partly with carbonic acid air, and partly with the required liquor, and then shaking it for a minute or two, &c. But the best way of performing this impregnation, is by means of a well-contrived machine, which has been long in use, and is generally known under the name of Dr. Nooth's glass apparatus, for making artificial mineral waters. There is, however, a contrivance for impregnating

pregnating water with an incomparably greater quantity of carbonic acid gas, than that which can be accomplifhed in Dr. Nooth's apparatus. But this contrivance is kept a fecret by the inventor, though the water, fo highly impregnated by him, may be had in London at a moderate price.

The application of factitious airs to the external parts of the body, may be performed with the utmoft facility. The aperture of a tube, which proceeds from the veffel in which the gas is generated, may be directed towards the part which is affected; a bladder full of the required gas may be gradually preffed, fo as to throw a ftream of the gas upon it; the part itfelf, as far as it is practicable, may be introduced into a veffel full of the required air; or, laftly, a fmall glafs funnel, with a bladder faftened to its fmall end, and filled with the required elaftic fluid, may be applied over the part, with the edge of its large aperture clofe to the fkin, fo as to prevent the efcape of the gas into the circumambient air.

THE medical application of factitious airs, and the effects which have thereby been produced, are as yet labouring under all the viciffitudes of truth and exaggeration, of accuracy and mifapplication, of fhort experience and uncertainty. The anxiety of fome perfons, the ignorance of others, the defire of fame, the love of intereft, and the fear of dangerous innovations, have alternately operated in favour and againft the adminiftration of the elaftic fluids for the alleviation of diforders incident to the human body. In the conflict of fuch oppofite powers, it is difficult to feparate truth from exaggeration and error; it is impoffible to afcertain the precife limits of their ufe and efficacy.

NOTWITHSTANDING thofe weighty objections, I have endeavoured to collect, to examine, and to methodize all the ufeful information which I could procure relatively to the fubject, in hopes that a comprehenfive view of it might promote the ufe, and in great meafure prevent the abufe,

abufe, of a new clafs of remedies, which have all the appearance of proving very advantageous to mankind.

IN the ufe of oxygen air we have a fingular ftimulus, which admits of its being rendered more or lefs active by dilution with various proportions of common air. In its pure, or nearly pure, ftate, it is a powerful exciter of fufpended animation; and when diluted with a confiderable quantity of common air, it is a gentle ftimulus, which, by invigorating the various parts of the animal body, by communicating firmnefs to the folids, and energy to the fluids, does frequently obliterate the caufes of morbid habits.

THE ufe of azotic gas, and of the various fpecies of hydrogen gas, produces a diminution of the irritability of the animal fibre to any degree, and hence it becomes ufeful in a variety of thofe diforders, which depend on an increafed irritability, fuch as inflammations, coughs, fpafms, &c.

In the ufe of the carbonic acid gas we have a powerful antifeptic, and in certain cafes a folvent of confiderable efficacy.

The ufe of pure oxygen air is confined to the purpofe of exciting the dormant powers of fufpended animation, and it is, therefore, to be adminiftered to children born apparently dead, or overlaid; to perfons fuffocated by drowning, by fteam of charcoal, by foul air, &c. whenever the circumftances of the cafe may indicate a poffibility of recovery.

Those cafes excepted, the refpiration of pure, or nearly pure, oxygen air, is almoft always attended with unfavourable fymptoms, fuch as a preternatural heat, efpecially about the region of the lungs; a quickened and feverifh pulfation; inflammation, &c. And thofe fymptoms come on after a fhorter or longer ufe of the oxygen air, according to the particular conftitution of the experimenter, and the purity of the gas.

But

BUT when the oxygen is diluted with much common air, *viz.* in the proportion of one to eight, and even as far as one to twenty, it then is a safe and very useful remedy, whose principal action consists in giving tone, elasticity, and consistence to the fluid as well as to the solid parts of the body, and of course it promotes all the natural consequences of those effects, *viz.* it quickens languid circulation, it strengthens the organs of digestion, promotes secretions, invigorates debilitated habits, and it assists nature in throwing off bad humours, and other lurking causes of diseases.

IT has been observed, that some individuals can bear a much greater proportion of oxygen than others, which is analogous to the various dispositions for all other applications. Thus a certain quantity of any remedy will act powerfully on some persons, whilst it will not be even felt by others. Thus also a certain quantity of food produces strength and cheerfulness in some individuals, whilst it produces sickness and indigestion

indigeſtion in others. It therefore becomes neceſſary, in the application of this remedy, to regulate the proportion of the two elaſtic fluids agreeably to the conſtitution of the patient, which may be eaſily accompliſhed by means of a very few trials.

IN the diluted ſtate, the oxygen air is adminiſtered by letting the patient breathe it for five or ten minutes once or twice a day. It might probably prove more efficacious, if it were breathed in a more diluted ſtate for a longer time; but the preceding mode has undoubtedly been attended with ſalutary effects.

HOWEVER ſlight this application may appear, however ſmall the unuſual quantity of oxygen which is thus introduced may be, the effects have been proved and confirmed by a variety of experiments and medical caſes. But independent of the experimental proof, the improbability of the effect will diſappear if it be conſidered; that the lungs of moſt perſons, and eſpecially of thoſe who labour under certain diſeaſes, are almoſt immediately relieved or affected by the tranſition from the air of one place to that of another; as by their going out of town, or even out of the houſe; and yet, as has been already obſerved, the difference between the air of a town and the air of the country, or of that of a houſe and of the external air, is ſo very trifling as hardly ever to be diſtinguiſhed by the eudiometer. Extremely minute, and almoſt inconceivably ſmall, quantities of matter can act with wonderful efficacy, when they are introduced into the circulation of the blood. The inoculation of the ſmall-pox, and the experiments with poiſons, furniſh ſufficient confirmation of this obſervation.

By breathing a mixture of common and oxygen air, even when the latter does not exceed one-eighth part of the former, for about ten or fifteen minutes, the pulſe is generally quickened of a few ſtrokes, but it is almoſt always made ſtronger. The lungs, during the operation, are ſeldom ſenſibly affected; but on leaving off the mixed airs, and

and returning to the atmofpherical air, a degree of tightnefs is frequently felt on the cheft, which, however, gradually goes off and vanifhes after a few minutes time.

WHEN debilitated habits breathe the diluted oxygen air for about a quarter of an hour once or twice a day, the improvement of their health is hardly ever confpicuous in lefs time than a week or a fortnight; but after that period, they will find their ftrength, their appetite, their digeftion, their circulation, and other functions, fenfibly improved; and this improvement goes on progreffively to a greater or lefs degree, according as age, local indifpofitions, times of the year, and other circumftances may allow.

THE mixtures of common air with azotic gas, or of common air and any fpecies of hydrogen gas, are commonly denominated *reduced atmofpheres*; for, in fact, they contain a fmaller quantity of refpirable fluid, than is contained in an equal quantity of of common air. The principal effect of those reduced atmospheres, is to diminish the irritability of the parts subservient to respiration, and indeed of the whole body; for which reason they are successfully administered in inflammations of the lungs, in spasmodic coughs, and in all the disorders that are nearly allied to those.

Much caution must be used in the administration of those reduced atmospheres, as some of them are productive of alarming symptoms. The mixture of azotic and common air, in which the former should never be more than a quarter of the latter, is the least dangerous, and at the same time the least efficacious. The same thing may be said of the mixture of common air with the mild sort of hydrogen gas, *viz.* that which is produced from iron and diluted vitriolic acid, or by passing the steam of water over red hot iron, excepting that it is rather more efficacious than the preceding. But the hydrocarbonate is much more powerful and more dangerous than any

any of the abovementioned gasses, especially when fresh made. It should, in general, be mixed with about twenty or thirty times its bulk of common air, unless some particular case may seem to require a greater proportion of the dangerous gas. For most purposes it will suffice to breathe it for about five minutes a day.

THE breathing of the diluted hydrocarbonate is attended with a diminution of sensibility, especially about the chest, and this effect is frequently so great, that some persons have expressed it by saying, that they felt as if they had no lungs at all, even when they had been a few minutes before in excruciating pains. But this diminution of sensibility is almost always accompanied with vertigo or giddiness, with a lowering of the pulse, and with faintness. It must be particularly remarked, that though those symptoms in general come on immediately after the operation, yet sometimes they return once, and even twice, more in the course of the day. When the breathing of
reduced

reduced atmofpheres proves very troublefome, it may be interrupted for a few minutes.

Its great power in checking irritability and fenfibility, feems to render the diluted hydrocarbonate applicable to fome diforders that have hitherto eluded all medical application; and as one of the moft likely to be relieved by this treatment, I fhall mention the hydrophobia, or madnefs which is occafioned by the bite of mad dogs, or other mad-animals *.

A REDUCED atmofphere, capable of diminifhing in fome degree the irritation of the lungs in inflammations, coughs, and

* I have been told, and have read, though I cannot at prefent recollect where, that the ufe of opium, and likewife that the fufpenfion of animation for a time by accidental drowning, have actually cured the hydrophobia in two or three cafes. If this be true, the probability of the hydrocarbonate proving beneficial, is thereby much increafed.

certain

certain fpecies of afthma, has been expeditioufly formed by mixing the vapour of vitriolic ether with common air. For this purpofe the patient needs only hold a fmall phial of ether open near his mouth, for about an hour at a time or longer, by which means the vapour of the ether mixes with the air that enters the lungs in the ufual courfe of refpiration, and converts it into an inflammable, or rather an explofive, aerial fluid *. For this purpofe it has been found ufeful to mix fome powdered leaves of hemloc *(cicuta)* with the ether. The ether *(viz.* about a quarter of an ounce of it) may alfo be put in a common tea-pot, and the mouth may be applied to the fpout of it, fo as to draw the air through it, and through the vapour of the ether.

THE carbonic acid gas has been longer in ufe as a medicine than any other facti-

* If three or four drops of ether be fhook in a phial full of common air, and if afterwards the aperture of the phial be prefented to the flame of a candle, the air in it will explode like a mixture of common air and hydrogen gas.

tious

tious aerial fluid. Much has been done, and much has been written, relatively to it. But the useful result of those experiments and investigations will be found condensed in the following few paragraphs.

In putrid fevers the free use of carbonic acid gas has been of considerable use, whenever the urgency of the case has not been very great, *viz.* when time was allowed for the gas to operate upon the morbid matter; and when the distention of the bowels was not so great as to prevent the free use of the gas.

In the scurvy this gas has been of considerable use in the beginning of the disorder, rather more than in an advanced state of it. But the use of vegetables, of sugar, and of other substances that contain it in great abundance, are acknowledged to be useful in all states of that disorder. Experience likewise informs us, that in the use of carbonic acid gas we are not to expect an unlimited antiseptic, nor a perfect solvent of the

the stone in the urinary bladder; yet its use in putrid cases, and in some diseases of the bladder or kidnies, is attended with considerable benefit.

The external application of carbonic acid gas to sores and ulcers of every sort, is unquestionably very useful.

After a careful consideration of the preceding general and comprehensive prospect of the medicinal use and efficacy of the aerial fluids, we may easily regulate the measure of our hopes by the standard of reason and experience. The idea of finding in them a remedy, capable of curing consumptions in all their stages, must be laid aside; and the hope of healing all sorts of internal ulcers will naturally vanish. The use of reduced atmospheres does undoubtedly diminish the irritability of the fibre, and a diminution of irritability favours the healing of certain ulcers, but by no means of them all; nay, in some cases it will even produce the contrary effect. The use of oxygen

oxygen air has been found advantageous in many of thofe diforders that are called nervous, and it has undoubtedly ftrengthened and invigorated feveral debilitated or emaciated habits; but it would be abfurd to expect that it fhould prove beneficial in all cafes of emaciation and debility, fince thofe vifible effects are often produced by caufes that may be rather fomented than checked by the ufe of oxygen air.

In moft of the diforders incident to the human body, the various concurring circumftances are far from being known to their full extent; hence theory may fuggeft, but experience muft prove the ufe of certain practices. Improvements and difcoveries may be generally urged and expected; but where theory and experience are filent, we have no warrantable guide to affift us in the inveftigation of new properties and new applications.

CHAPTER VII.

Of the particular Administration of aerial Fluids in different Disorders.

AFTER a general idea of the application of factitious airs by way of remedies to the human body, it will be necessary to state those modes of treatment, which experience or analogy shew to be the most efficacious in particular diseases. But this statement cannot be attended either with great accuracy of description, or with extensive information concerning the phænomena, that are really produced by the factitious airs in all cases. The various nature of individuals, the imperfect accounts of several cases, and the frequent administration of other medicines in conjunction with the aerial fluids, limit for the present the attainment of the abovementioned objects.

ALL

ALL that the practitioner may expect to derive from the present chapter is, a guide or indication for the commencement of the application, a general view of the principal effects that are produced by the particular administrations, and a warning against mistakes. But with respect to the continuation, or suspension, or alteration, of the treatment, he can only be instructed by a careful observation of the phænomena which take place in the course of the application.

I SHALL forbear mentioning other medicines that may be proper to be administered at the same time with the gasses, as these must be left to the judgment of professional gentlemen. But I would strongly recommend to administer them as sparingly as the nature of the case can possibly admit of; being persuaded, that the good effects of the aerial fluids is frequently counteracted by the action of other medicines.

N. B. The diseases in the following pages are arranged in alphabetical order.

Animation suspended.

IN cafes of this fort, whether they be occafioned by drowning, by noxious vapours, or by any other caufe of the like nature, the oxygen air fhould be adminiftered pure, or nearly fo. The wooden pipe of a large bladder full of it muft be introduced into the mouth of the fubject, the lips muft be preffed upon the faid pipe, and the noftrils muft be ftopped by the hands of an affiftant. Then by preffing the bladder, the oxygen air muft be forced into the lungs, as much as poffible, for about eight or ten feconds, after which the mouth and noftrils being unftopped, without removing the pipe of the bladder, the cheft about the region of the lungs muft be preffed gently; then the bladder being applied as before, the oxygen air is forced again into the lungs, and fo on; continuing a fort of forced and artificial refpiration for about a quarter of an hour at leaft, if no figns of life

FACTITIOUS AIRS. 117

life appear before that time *. But as foon as any natural or fpontaneous movements are perceived, the preffing of the region of the lungs may be difcontinued, and the bladder, &c. muft be removed; for in that ftate a free ventilation of the ambient air will be found fufficient to reftore life.

This treatment fhould be accompanied with the communication of a gentle warmth, and perhaps with friction to the hands and feet. But care muft be taken to do what is juft neceffary, and not too much; for in the attempts to reftore animation, the ftimuli and other applications are frequently carried fo far, as to deftroy that laft fpark of life, which they were intended to revive.

In cafes of children born apparently dead, or ftrangled in laborious parturition, &c. the ufe of oxygen air cannot be too forcibly recommended. The application is eafy and highly promifing. Independent of

* Several bladders full of oxygen air fhould be kept in readinefs, for a fingle bladder will be foon exhaufted.

the

the experiments that have been made on brutes, I know of a cafe, in which a child born apparently dead, was brought to life merely by forcing oxygen air into his lungs, whilft he was held before the fire.

Afthma.

I FIND many creditable accounts of this diforder having been relieved, and fometimes perfectly cured, by the ufe of diluted oxygen air in fome cafes, and by the ufe of reduced atmofpheres and the vapour of ether in other cafes.

IT would be abfurd to imagine, that either of thofe treatments may be indifcriminately applied to the very fame fpecies of afthma; but the diftinctions are not clearly ftated in all the accounts of the cafes. It appears, however, that in a plethoric afthma, and when the diforder is attended with confiderable pain, hard cough, and inflammatory fymptoms, the reduced atmofpheres muft be adminiftered.

IN

In thofe cafes the patient may be directed to breathe daily fixteen quarts of common air, with four quarts of hydrogen, obtained from iron and diluted vitriolic acid, or, which is better, from the vapour of water and red-hot iron. But fhould this mixture of elaftic fluids prove ineffectual in a day or two, then a mixture of one pint of hydrocarbonate and thirty pints of common air, may be ufed inftead of it; and the ftrength of this mixture may be increafed according to circumftances. If in breathing the diluted hydrocarbonate, giddinefs fhould come on, the patient muft be defired to intermit the operation; refting, that is breathing the ambient air, for a few minutes, and then to refume the inhalation of the diluted hydrocarbonate. Thus the operation may be intermitted three or four times.

The breathing of the vapour of ether, after the manner defcribed in the preceding chapter, has been found to afford confiderable alleviation of the pain and oppreffion in thofe cafes.

IN nervous afthma, and efpecially in debilitated habits, the oxygen air may be adminiftered; and it will be proper to begin by inhaling daily eight quarts of common, with two quarts of oxygen, air, extracted from manganefe by means of heat. The quantity of oxygen may be increafed, in cafe the abovementioned proportion fhould be found ineffectual; and it is remarkable, that in this fpecies of difeafe the patients can fometimes bear a great quantity of oxygen without any material effect.

IN all cafes of afthma, the effects of the application of factitious airs may be perceived in the courfe of four or five days; but the accomplifhment of the cure will frequently require fix weeks time, or longer.

Cancer.

THE ftubborn nature of a cancer, and its dreadful confequences, render every hint, that promifes an alleviation of its effects, extremely interefting.

THE

The elaftic fluids have been repeatedly applied to cafes of this fort, and fuch applications have been attended with confiderable advantage. I do not find any authentic account of a cancer having been completely cured by the ufe of factitious airs. But certain it is, that in a variety of cafes the pain has been confiderably diminifhed, the fœtor as well as the bad afpect of the ulcer, have been almoft entirely removed, and the whole habit of body has been confiderably improved, fo that the patients have thereby been enabled to have comfortable nights, more cheerful countenances, &c,

Those good effects have been produced by the external application of carbonic acid gas to the ulcer, and the inhalation of diluted oxygen air. Both thofe elaftic fluids muft be adminiftered daily for weeks, or as long as the indications of the cafe may afford a hope of melioration. The manner of applying the carbonic acid gas has been already

already defcribed *; as for the continuance of the application, an hour a day is by no means too much, and it would be better if fuch an application were repeated two or three times in the courfe of each day. With refpect to the oxygen, two or at moft three quarts of it, with about fourteen or fixteen quarts of common air, may be fufficient for each daily inhalation.

Catarrh.

In colds and defluxions, efpecially when accompanied with tightnefs about the region of the lungs, and a hard cough, much and almoft inftantaneous relief has been frequently obtained by breathing a mixture of about four quarts of hydrogen and twenty quarts of common air. There is no need of breathing this quantity at once. It will hardly ever be neceffary to repeat this application longer than the third day. The breathing of the vapour of ether in the

* See chap. VI. p. 99.

manner already defcribed, will anfwer nearly as well as the above-mentioned mixture of elaftic fluids, and it has the advantage of being a much eafier application, fince it requires no particular apparatus *.

Chlorofis.

THE adminiftration of diluted oxygen air has proved beneficial in difeafes of this kind, perhaps more often than in any other diforder, as is proved beyond a doubt by feveral authentic cafes. The palenefs, the debility, the palpitation, the fever, the depraved appetite, and the other bad fymptoms which accompany this diforder, generally begin to diminifh in about four or five days, and a complete cure is often accomplifhed in about fix weeks time.

THE daily inhalation of one quart of oxygen, and ten or twelve quarts of common air, may fuffice for the beginning. But it

* See chap. VI. p. 110.

is

is to be remarked, that chlorotic patients are senfible of the least excefs in the proportion of oxygen, so that sometimes they are more hurt than benefited by it, unless such a quantity of it be administered as may be just necessary; and this quantity can be shewn only by a careful obfervation of the effects which take place. The lungs will be enabled to bear the stimulus of oxygen air every day better and better.

Confumption.

THE various states of confumption, or *phthifis pulmonaris*, its different caufes, and the difficulty of difcerning a real phthifis from certain other diforders, render the treatment of this difeafe frequently doubtful and perplexing. But its stubborn nature, and the frequency of the difeafe, demand the utmoft attention, and all the afliftance which philofophy can fuggeft, and the medical art can apply. We shall therefore endeavour to state, how far the ufe of factitious airs has been found ufeful or promifing in cafes of this nature.

It has been said on one side, that the factitious airs have the power of arresting the progress of consumption, and often of accomplishing a perfect cure; but on the other hand it has been asserted, that they have never afforded any permanent benefit, and that they have often produced manifest harm. It appears, however, from a disinterested examination of the cases, and from the testimony of patients as well as of practitioners, that both those assertions imply a considerable degree of exaggeration. The result of this examination will be found condensed in the following paragraphs.

The diluted hydrocarbonate is the only one, or at least the principal aerial fluid that has been successfully administered in cases of phthisis; and it has generally afforded a sensible and almost immediate relief, by abating the hectic fever, by diminishing sensibility, by promoting sleep, and by reducing the quantity of expectoration.

But the use of hydrocarbonate is always attended with a diminution of strength. Hence,

Hence, when the patients are very feeble, which is generally the cafe in an advanced ftate of the diforder, the difadvantage which arifes from the diminution of ftrength, is greater than the advantage which arifes from the other good effects of the hydrocarbonate. When the patients, therefore, are too far gone, the ufe of the hydrocarbonate produces an apparent but not a real melioration.

It is on the fame account that this elaftic fluid cannot be adminiftered to patients, that labour under great weaknefs of the digeftive organs. In fuch cafes the vapour of ether is, perhaps, the only elaftic fluid that may be tried with fafety; and the ufe of it is attended with at leaft a temporary relief.

There are two or three cafes of real phthifis creditably related, where a perfect cure feems to have been performed; though in a great many others the application of aerial fluids proved evidently ufelefs. But though

though from thofe few fuccefsful cafes no great expectations can be derived, yet in a difeafe where no remedy has ever been found efficacious, furely it is not improper to try an application which at leaft affords a ray of hope.

The quantity of diluted hydrocarbonate, which may be adminiftered daily, is various, according to the conftitution of the patient. It is proper, however, to begin by adminiftering one pint of hydrocarbonate with between twenty and thirty pints of common air; and the quantity of the former may, in procefs of time, be increafed conformably to the effects. In breathing this quantity of elaftic fluid, it will be proper to let the patient reft four or five times, or in fhort whenever any giddinefs happens to come on; for this giddinefs or vertigo generally goes off in two or three minutes, after which the patient may again apply his mouth to the bag or veffel which contains the diluted hydrocarbonate.

It

It will be found, that cuſtom habituates the lungs to bear the hydrocarbonate, in an increaſed proportion, as far as a certain limit. Thus the ſame patient who at firſt was made vertiginous by a quart of hydrocarbonate, diluted with twenty quarts of common air, will, in proceſs of time, be hardly affected by the double of that quantity.

The inhalation of the vapour of ether, as alſo of other ſorts of reduced atmoſpheres, ſuch as a mixture of azote and common air, of carbonic acid gas and common air, of hydrogen and common air, have been of partial uſe; however, the mixture laſt-mentioned ſeems to have proved more beneficial than any of the reſt. This ſort of reduced atmoſphere muſt be adminiſtered more freely than the diluted hydrocarbonate. The vapour of ether may be inhaled with the utmoſt facility, as no apparatus is required for it, and it will be found at leaſt of temporary uſe for allaying the cough, the pain, &c.

Though

Though the use of reduced atmospheres be more promising in cases of incipient phthisis, yet that application should not be neglected in any state of the disorder; since the elastic fluids are the only remedies which can be applied immediately to the part affected.

Of the various species of phthisis pulmonaris, two only, *viz.* the chlorotic and the syphilitic, seem to require a different treatment, and I find a few cases in which syphilitic ulcers in the lungs are said to have been cured by the use of diluted oxygen, which was breathed once a day; but this treatment was accompanied with mercurial and other medicines; which, however, when administered by themselves, had produced no good effect.

The inhalation of carbonic acid gas is said to have proved beneficial in hectic disorders, but I do not know how far this practice may be safe or useful, as I do not find any very particular information concerning it.

Coughs.

OF the various species of cough, those which originate from catarrh and from phthisis have been already mentioned under those articles, to which the reader is referred. But with respect to the application of factitious airs to other species of cough, I do not find much authentic information, and of course must leave it for future investigation.

Debility.

AN universal debility is not unfrequently met with amongst persons of all ages, and especially among women. It is sometimes the unconquerable effect of former disorders that are subdued, or of lurking and invisible causes. Whatever its origin may be, the symptoms it produces are numerous and often of the utmost consequence. It produces paleness, emaciation, difficulty of breathing, palpitation, indigestion, loss of sleep, frequent cough, swellings of the extremities,

tremities, weakness of fight, loss of voice, suppression of the usual evacuations, &c. Those symptoms, of which a greater or less number is to be observed in the same individual, are, at first, the consequence of the debility, but they soon become the fomenters of that very languor, and consequently of each other.

WHEN the disease, which produces the languor, is present and known, I need hardly mention that the removal of that cause should be the first object of the practitioner. But when that is not the case, diluted oxygen air may be administered with great hopes of success; for such treatment has been found beneficial in a great many cases of this sort, and wonderful cures have been performed where no other remedy was found efficacious. The improvement is perceived sooner or later, according to the nature of the cases; but, generally speaking, it becomes manifest in about a fortnight or three weeks time. It operates by strengthening and improving the whole habit. The pulse

pulse becomes stronger, the aspect acquires colour, the lassitude after exercise goes off gradually, the appetite is improved, and the rest of the symptoms disappear gradually.

It has been repeatedly observed, that in cases of this sort the quantity of oxygen must be nicely regulated by the strength of the patient. If too small a quantity be administered, little or no improvement will be obtained; but if the quantity be too great, the effects will be hurtful, and some of the bad symptoms are thereby increased. A few days experience will soon indicate the proper dose of oxygen. But I would recommend to begin by giving one quart of oxygen, with twelve or fifteen quarts of common air, per day, and to increase or diminish the quantity according to circumstances. It must, however, be remembered, that when a sensible improvement becomes manifest, more advantage is to be derived from a moderate dose regularly administered, than from an increased proportion of oxygen.

FACTITIOUS AIRS. 133

I do not find that the leaving off of this application has ever been productive of any harm, but at all events it may be not improper to relinquish it by degrees, *viz.* by diminishing the quantity of oxygen, and intermitting the application by the interval of a day or two.

Digestion impaired, or Dyspepsia.

WE have not a clear account of the various species of dyspepsia to which the aerial fluids have been applied, nor indeed have they been tried in a great variety of cases. But upon the whole it appears, that when debility is the cause, and especially when it is accompanied with what is called a nervous head-ach, the inhalation of diluted oxygen air has been of singular use, and the disorder has been frequently removed in a short time.

If the impaired digestion be accompanied with other symptoms of debility besides the head-ach, the administration of oxygen may be regulated agreeably to what has been mentioned in the preceding article, otherwise a greater proportion of it may be administered, as about four or five, or six pints of it, with between twenty and thirty of common air. In cases of this sort, the good effects of the oxygen may be perceived in the course of a few days.

THIS treatment has proved peculiarly beneficial to such persons as have contracted a weakness of digestion, from having been confined in the foul air of workshops, counting-houses, &c.

I FIND, likewise, the case of a man who had been afflicted for upwards of five years with heart-burn, flatulence, lowness of spirits, and coldness of the extremities, which seemed to indicate a bad digestion, and who was perfectly cured by the inhalation of diluted oxygen, and by drinking water impregnated with carbonic acid gas, together with some salt of steel.

Dropsy.

Dropsy.

IN a variety of dropsical cases the inhalation of diluted oxygen air has been attended with success, and this success has several times amounted to a complete cure. This treatment seems to be more efficacious in an incipient dropsy, and when the disorder is confined to the extremities, than in other states of it. Yet I find a remarkable case of hydrothorax which was effectually cured, though a similar one was not attended with the same effect; and likewise a case of water in the head of a boy of thirteen years, which is said to have been partially removed by the inhalation of diluted oxygen air.

ONE quart of oxygen, and about fifteen of common air, per day, may be sufficient for the beginning; but the proportion of oxygen must be increased in the course of three or four days (provided no bad effects ensue) to two quarts; and soon after it will be be proper to double the quantity both of oxygen and of common air.

THIS regimen muft be perfifted in for weeks and months, according to the nature of the cafe; and fhould any inflammatory fymptoms appear in the courfe of this application, the inhalation of the diluted oxygen may, in that cafe, be fufpended or moderated for two or three days.

Eruptions.

I FIND a few cafes of fcorbutic eruptions on the face, as alfo on other parts of the body, in which a complete cure was accomplifhed by the daily inhalation of diluted oxygen air.

IN cafes of tumors and eruptions, which derive their origin from debility and a poor or thin ftate of the blood, the like treatment has been found beneficial.

Two

Two or three pints of oxygen, with about ten times that quantity of common air, per day, is sufficient for the beginning; but in cafes of this fort, the proportion of oxygen fhould not be much increafed.

I NEED hardly add, that in fuch cafes it is proper to continue the ufual dreffings of the parts affected, the means of keeping the body gently open, &c.

Fevers.

I DO not know whether the factitious airs have been tried with fuccefs in other forts of fever, befides the putrid and the hectic. With refpect to the latter, the reader may confult what has already been faid under the articles of *chlorofis* and *confumption*. But in putrid fevers the carbonic acid gas is generally allowed to be an ufeful remedy; yet the ufe of it has not proved fo generally beneficial as it was at firft believed.

THIS

This gas is applied internally, not to the lungs but to the ſtomach and inteſtines, in three different manners, *viz.* by way of clyſters, either in the aerial form or in combination with water; by way of drink when combined with water; and, laſtly, by giving through the mouth ſuch ſubſtances as contain carbonic acid gas in abundance, that is liquors in a ſtate of fermentation, certain fruits, &c.

When a large quantity of it is given either in the aerial form or in combination with water, the abdomen is frequently diſtended by it; for though this gas is pretty eaſily imbibed by animal fluids, the fluids which it uſually meets with in the ſtomach, &c. are ſeldom capable of abſorbing more than a moderate quantity of it. However, the diſtention of the abdomen is not ſo very detrimental, but it may be ſupported to a certain degree.

What ſeems to render the carbonic acid gas not ſo efficacious in caſes of putrid fevers,

fevers, as from its usual properties one might be led to expect, is the difficulty of its insinuation into the vascular system of the whole body. The lacteals imbibe it in small quantity, and the difficulty becomes greater in certain states of the disorder; whenever, therefore, the disease is not in a very alarming state, *viz.* so as to give time for the insinuation of the carbonic acid gas into the fluids of the body, then more benefit is to be expected from it. There are, however, some cases in record, where the free use of carbonic acid gas proved efficacious in the worst state of putrid diseases; and I do not find that it was ever attended with noxious effects.

Of the various substances which are administered in putrid diseases, on account of the carbonic acid gas which they contain, the following are the principal ones, *viz.* effervescing alkaline and acid mixtures, consisting of a solution of salt of tartar, to which lemon juice, or diluted vitriolic acid, or diluted nitrous acid, is added the moment

ment before it is to be drank; fweetwort, or an infufion of malt, yeaft, and certain acidulous fruits, fuch as oranges, lemons, &c.

I SHALL not attempt to define the circumftances in which one or other of thofe articles may be preferable, nor is it neceffary to limit the dofes. The circumftances of fuch cafes being very numerous and diverfified, muft be left to the fkill of the attending practitioners. If the carbonic acid gas be adminiftered in the aerial form, the quantity of it can hardly be too great, provided it does not diftend the abdomen too much; but if the gas be given in combination with other fubftances, the quantity of thofe other fubftances muft be limited, not by the quantity of carbonic acid gas that may be contained in them, but by their other properties, which muft be proportionate to the ftate of the patient.

IT has been propofed (not without expectations of fuccefs, though with difficulty of of execution) to introduce, in certain cafes, the whole body of the patient, the mouth excepted, into a veffel full of carbonic acid gas; for as this gas is abforbed by the pores of the fkin, a greater quantity of it might thereby be imbibed.

Head-Ach.

THE various origin of this diforder, and the fmall number of cafes that are circumftantially related, prevent our forming a comprehenfive idea of the ufe of factitious airs in cafes of this fort. The inhalation of diluted oxygen air, has fometimes been of ufe in what is commonly called nervous head-ach; and it appears that in fuch cafes, a very great proportion of oxygen air has been adminiftered, even as much as five or fix gallons per day. I would not, however, recommend fo free a ufe of it.

IN head-achs that arife from a weak digeftion, the inhalation of diluted oxygen air is an uieful remedy. See the article *Digeftion*.

Hæmoptyfis,

Hæmoptyſis, or Spitting of Blood.

I FIND a few caſes of this diſorder, in which the adminiſtration of hydrocarbonate gas was adminiſtered with good effect. The account of the moſt remarkable one will be found in the next chapter.

Ophthalmia.

IN caſes of ophthalmia, and weakneſs of ſight, when accompanied with a relaxed habit of body, the inhalation of diluted oxygen air has been of ſingular uſe. About two quarts (when a ſmaller quantity has proved ineffectual) of oxygen air, with about fifteen of common air, is a doſe ſufficient for each day.

Phthiſis Pulmonalis.

See *Conſumption*.

Paralyſis.

Paralysis.

I FIND a few cases of that species of paralysis which is occasioned by preparations of lead, the *colica pictonum*, where the inhalation of diluted oxygen air proved beneficial. Three or four pints of it, with about thirty pints of common air, is a dose sufficient for each day.

Scurvy.

THE use of carbonic acid gas has long been considered as a powerful remedy in scorbutic disorders; and certain it is, that when the disorder is not too far advanced, a perfect cure may be generally expected from it; and even in cases of the worst sort, the free use of this gas has frequently accomplished a perfect recovery.

ALL the various ways of administering this gas, which have been mentioned for the cure of putrid fevers (See the article *Fevers)*

Fevers) are applicable to this fort of diforder.

Since much has been written concerning the fcurvy, and fince the methods of adminiftering fuch fubftances as contain abundance of carbonic acid gas, are generally known and fuccefsfully adminiftered, I fhall not detain the reader with long extracts from more able writers; but fhall only add, that, whilft the carbonic acid gas is applied to the ftomach and inteftines, a moderate dofe of diluted oxygen air fhould be applied to the lungs by the way of refpiration; for whilft the former acts as an antifeptic, and corrects the putrid tendency, the latter gives energy and vigor to the fibre, and enables the body to throw off the morbid humours with greater quicknefs.

Stone in the Bladder, &c.

WATER impregnated with carbonic acid gas, has been long known to afford relief in calculous complaints of the bladder and

and urinary paffage. But by the addition of a fixed alkali, the remedy has of late been rendered much more efficacious in cafes of the above-mentioned fort, and even when a large ftone has actually exifted in the bladder. I do not know how far this *acidulous foda water*, as it is commonly called, may operate by way of a folvent of a large ftone; but certain it is, that even in thofe cafes it affords confiderable relief, and it feems effectually to prevent the farther accumulation of the ftony matter, by diffolving the mucus as well as the fmall ftony concretions, and wafhing them off from the kidnies, ureters, bladder, &c. It is, therefore, given in all complaints that originate from a thickening of, or depofition of grofs matter by, the urine in the above-mentioned parts, fuch as ftrangury, pain in voiding the urine, ulceration of the parts, &c.

For this purpofe one ounce of foda is diffolved in four or five pints of rain, or of boiled foft, water; and the folution is then impregnated, as much as poffible, with carbonic

carbonic acid gas *. Of this water, a pint a day is the quantity ufually given for the above-mentioned diforders, and it is to be drank not all at once, but at three different times, *viz.* morning, noon, and night.

INDEPENDENT of thofe difeafes, the acidulous foda water is fuccefsfully adminiftered in fcorbutic cafes, bilious complaints, weaknefs of the digeftive organs, fome nervous affections, &c. but in thofe the proportion of the alkali, as well as the daily allowance, muft be diminifhed according to the circumftances of the cafe.

Swellings.

I FIND one cafe only of a white fwelling of the knee recorded, in which a per-

* In Dr. Nooth's glafs apparatus for impregnating water with carbonic acid gas, the quantity of gas that can be thrown into it is very moderate, yet efficacious; but the foda water which is now prepared and fold in London by a Mr. Schweppe, contains an incomparably greater proportion of carbonic acid gas, and accordingly is much more efficacious.

fect

fect cure is said to have been accomplished by the daily inhalation of diluted oxygen air. As those swellings owe their origin in great measure to weakness of body, it is likely that the use of oxygen, which invigorates the animal fibre, may prove an useful remedy.

THE like treatment is said to have been found sometimes useful in scrophulous tumours.

Ulcers.

THE factitious airs have been frequently administered in cases of ulcers on different parts of the body, and especially on the legs; but the indiscriminate and injudicious application, which seems evident in many cases, has been productive of equivocal effects. However, a careful examination of the particular circumstances shews, in agreement with the theory, that when the ulcers originate from a poor state of the blood, and a debilitated habit, the daily inhalation

inhalation of three or four pints of oxygen air, with about ten times that quantity of common air, is of fingular ufe; and by this means fome ulcers of the worft kind, *viz.* painful, fœtid, ftubborn, &c. and when they were accompanied with fcurfy eruptions over great part of the body, with want of appetite, &c. have been completely cured in about fix weeks time.

In ulcers of other fort the ftate of the patient, as alfo the origin of the diforder, muft be carefully attended to, and the elaftic fluids, when they may be thought ufeful, muft be adminiftered accordingly, otherwife they will produce more harm than good. In fact, I find a cafe of a fcrophulous ulcer, where the oxygen air proved detrimental; but a mixture of oxygen, hydrocarbonate, and common air, accomplifhed the cure. This cafe will be found in the next chapter.

In all cafes of ulcers, the external application of carbonic acid gas generally affords an alleyiation of pain, as well as of the fœtor, and a better difcharge.

CHAPTER VIII.

Medical Cafes in which Aerial Fluids were adminiftered.

A COLLECTION of medical cafes, in which the factitious airs have been adminiftered with great fuccefs, forms the content of the prefent chapter. Thofe cafes have been either extracted from other publications, or they have been communicated by intelligent friends; and they have been felected out of a great number, merely for the purpofe of fhewing the practical methods of adminiftering the artificial elaftic fluids. Such cafes, therefore, have been preferred, as by the variety of circumftances feemed more likely to manifeft the modes of applying, proportioning, varying, and fufpending suspending the administration of the factitious airs.

Convinced that the unskilful application of this new set of remedies has produced considerable harm, and has thrown a degree of discredit on the practice, I take the liberty of warning the practitioners against drawing hasty conclusions from a few crude, and, in all probability, ill-managed cases. For however skilled those gentlemen may be in other branches of physic, it is at least likely that in this new application their management of patients may not be generally correct; and of course the failure is not always to be attributed to the want of power in the aerial fluids.

Case I.

Communicated by Dr. J. Lind, of Windsor.

THE first time I applied the modified atmosphere as a remedy, was in the case of an officer of the Excise, who, during the

severe

severe weather of January, 1797, being much exposed to the cold in the exercise of his duty, had got a violent cough, which caused the rupture of a considerable blood vessel in his lungs, and this was soon followed by the symptoms of a rapid consumption. On the 25th of January he first applied to me, when I ordered him an infusion of roses, acidulated with vitriolic acid, and small doses of ipecacuanha, to stop the hœmoptoe; and for the cough and hectic fever I recommended him to breathe, several times in the course of the day, the vapour of vitriolic ether, in which the powdered leaves of *cicuta* were infused, after the manner recommended by Dr. Pearson, of Birmingham. The benefit which he received from this application was really remarkable, for after not more than four or five days, almost all the bad symptoms were wonderfully diminished; yet, finding that he got hardly any sleep at night, and that he had been a bad sleeper for above a year, I made him inhale about a quart of hydrocarbonate gas, diluted with about fourteen quarts

quarts of common air, at bed-time, which procured him an uninterrupted night's reſt, ſuch as he had not experienced for many months before. He continued to follow the ſame courſe till the 20th of February, when his health being perfectly reſtored, he returned to his duty.—N. B. When he inhaled the diluted hydrocarbonate, he drew it in over the ſteam of hot water, with the inhaler which I have conſtructed on the plan of Mr. Watt's refrigeratory *.

CASE II.

Communicated by the ſame.

I TRIED the diluted hydrocarbonate likewiſe with ſucceſs, in an inflammation of the lungs. The patient was a man of ſixty-ſix years of age, labouring under an inflammation of the lungs, but attended

* This inhaler is made of japanned tin, and being filled with hot water, is interpoſed between the bag or receiver of air and the mouth of the patient, ſo that the air is heated by the vapour of hot water in paſſing through it.

with

with fo fmall a pulfe that bleeding did not appear advifeable. I therefore directed him to breathe the hydrocarbonate gas, diluted nearly in the proportion mentioned in the preceding cafe, which he did every night, and occafionally whenever the pain returned. The effect of the modified air was immediate, and very remarkable, for not only the pain was removed, but he ufed to fay that the hydrocarbonate had deprived him of his body, and had left him only his head; fuch was the diminution of irritability which this gas is capable of producing.

The diforder vanifhed in a very fhort time: for in feven days from the commencement of the application, his health was perfectly reftored.

This cafe fhews that in inflammations of the lungs, when the pulfe is weak, which is fometimes the cafe, the reduced atmofpheres are, perhaps, the only application practicable.

CASE

CASE III.
Related by Dr. Carmichael.

Birmingham, March, 1795.

I. B. æt. 45, was attacked about four months since with difficulty of breathing, attended at times with pain under the sternum, and commonly with a sense of tightness of the thorax, frequent cough, with copious expectoration of a tough whitish fluid, pulse 96, body regular, appetite variable. He has seldom passed four and twenty hours without a material aggravation of all his symptoms. Was first attacked with this disorder six years ago, and has regularly suffered very severely from it every winter since that period; it has always left him about the beginning of May, and he has kept free from complaint during the summer and autumn months. He has tried many remedies, but never with more than very transitory relief.

FEBRUARY 14th, 1795, I directed him to inhale daily a mixture of hydrocarbonate and

and atmospheric air, in the proportion of 1 to 19.—15th, No sensible effects from the use of the hydrocarbonate; the strength of the mixture was therefore increased in the proportion of two to 18.—16th, No vertigo, nor any other sensible effect, produced by the use of the modified air. The proportion still farther increased to 4 to 18.—17th, Considerable vertigo produced by yesterday's dose, which returned at intervals, attended by head-ach during the day. Breathing much relieved, even during the act of inhaling the modified air, and has since continued tolerably easy. Slept better last night than he has been accustomed to do for some months.—22d, Hydrocarbonate continues to produce considerable giddiness; breathing, except some short intervals of slight return, continues much easier. Cough less frequent, expectoration much diminished. Continues to enjoy comfortable sleep.—27th, Had a considerable return of difficulty of breathing on the afternoon of the 25th, which, however, abated so much before his usual bed-time, as not to prevent him from paſſing the night comfortably. Cough infrequent, and rarely attended with expectoration. Has for ſome time paſt had no pain under his ſternum, and rarely any ſenſe of tightneſs of his thorax. —March 4th, He is in every reſpect ſo much better, that he intends to return to his uſual occupation (making moulds in a caſt-iron foundry) on the 9th inſt. Modified air continues to produce vertigo.— March 9th, He continued without any return of his complaint, and returned to his employment as he intended; but after working for a few hours only, he was obliged to deſiſt, by a return of the ſenſe of tightneſs on his thorax, and conſiderable difficulty of breathing. Breathing increaſed in difficulty towards evening, and ſtill continues, attended by frequent dry cough.— 13th, Continues to breathe with conſiderable difficulty, pulſe 100; ſleepleſs nights; cough more frequent, but now attended with conſiderable expectoration. —17th, Difficulty of breathing continued until yeſterday; has paſſed a better night than uſual;

ufual; and this morning finds himfelf much better.—20th, Breathing continues eafier; cough much lefs frequent, and quantity of expectoration diminifhed. Has flept for fome nights paft comfortably; pulfe 86. Modified air continues to produce confiderable vertigo.—29th, Continues uniformly to recover; his cough is very trifling, and he expectorates better; his ftrength is fo much improved, that he can ufe confiderable exercife without inconvenience. Sleeps uniformly well.—He returns to his work to-morrow, but for the prefent is to work within doors. He is of opinion that he is in every refpect equal to the undertaking.

CASE IV.

Related by the fame.

J. T. æt. 40. has for two years paft been affected, during the winter and fpring months, with cough and expectoration, and at times with pains in his breaft, accompanied with flight dyfpnœa. Thefe fymptoms,

toms, in general, left him during the summer months, and never at any time arose to such a degree as to prevent him from following his usual occupation. In the beginning of October last he was seized with pain on his side, cough, dyspnœa, and after some time with copious expectoration. He applied for my advice in the beginning of November. At that time he had an almost inceſſant cough, attended with copious expectoration; he complained of a sense of tightness across his thorax, and much dyspnœa on the slightest exertion; his pulse was in general from 110 to 120, his nights were restless, and attended with profuse perspirations, his body was irregular, his appetite much impaired, his frame much emaciated. I ordered for him, at different times, emetics, squills, ammoniacum, blisters, &c. but from none of them did he derive more than a very temporary relief.—November 27th, he began the use of the hydrocarbonate; I directed him at first to inhale a mixture containing a quart and a half of this species of factitious air, and

nineteen

nineteen of atmofpheric air. This quantity he ufed in about twenty minutes, breathing it for twenty feconds together, and then refting for one, two, or three minutes, according to the degree of vertigo produced. —28th, The vertigo produced by yefterday's inhalation was very fevere, and returned at intervals during the evening. He has paffed a much better night than ufual, and fays that the dyfpnœa and fenfe of ftricture on the thorax are much relieved. The quantity of hydrocarbonate diminifhed to one quart, diluted as above.—30th, Cough much relieved, fenfe of ftricture gone, dyfpnœa lefs troublefome on motion, has had better nights, and his perfpirations are lefs profufe; pulfe 106, appetite rather better.— December 7th, Cough evidently better, expectoration confiderably diminifhed, pulfe 95, body for fome days paft regular; breathing fo much improved that he can with eafe walk up ftairs to his chamber and undrefs himfelf, without return of difpnœa, which he could not before accomplifh without the greateft difficulty; fleeps better than

he

he has done for months paſt, perſpirations entirely left him, appetite mended.—15th, Continues to recover in every reſpect, has at times ſome return of tightneſs on his breaſt, but which is uniformly relieved or completely carried off by the hydrocarbonate. His countenance is evidently altered for the better, and he is of opinion that his ſtrength returns. Notwithſtanding that the modified air ſtill continues to produce conſiderable vertigo, I increaſed the quantity to two quarts, diluted as before.—27th, Cough very much relieved, expectorated matter reduced to one-third of its former quantity, pulſe from 84 to 90. He has evidently acquired fleſh, and he is of opinion that his ſtrength continues to improve.— January 6th, 1795, Cough rather more frequent, and attended with ſome degree of diſpnœa. On account of the ſeverity of the weather, which evidently affects him, I ordered him not to ſtir from home. At this time he began to breathe the modified air, of the ſtrength directed above, twice a day. —16th, Cough relieved, quantity of expectorated

torated matter much the same as reported on the 27th ult.; in other respects the same.—February 1st, On account of the unusual severity of the weather, no advance has been made since last report. Cough more variable, and at times attended with some degree of dispnœa, expectorated matter increased, he does not, however, emaciate. —12th, Cough much abated, quantity of expectoration reduced to one-fifth of its former quantity, his strength is so much recruited that it is with difficulty I can restrain him from returning to his occupation. In every respect he is much better.—March 1st, continues to gain strength, cough less frequent, and expectoration still diminishing in quantity, appetite good, sleeps well. As I could not prevail with him to remain longer at home, I advised him, before he returned to his usual occupation, to walk out a little daily.

He complied with my advice, and continued to gain ground till the 9th of that month, when in the evening he was seized with

with the ufual fymptoms of the influenza, an epidemic catarrhal infection, which at that time prevailed much in this place. The febrile fymptoms ran high, and were attended by frequent cough and confiderable pain on his fide; he complained alfo of fevere head-ach, and unufual langour; he was thirfty; his tongue was white, and his pulfe 110.—March 15th, febrile fymptoms continue; cough frequent, but now attended with increafed expectoration; pain of his fide lefs fevere; confiderable difpnœa on the flighteft motion; pulfe 115, fmall and weak. Until this attack he inhaled twice daily a gallon of hydrocarbonate, diluted with four times the quantity of atmofpheric air, but, as his ftrength wafted, it was found neceffary to leffen the quantity to one quart diluted as above.—March 20th, pain of his fide fomething eafier, but his cough is increafed in frequency, and his expectoration more copious. Reftlefs nights; no appetite; ftrength fo much impaired that, for the laft four days, he has not been able to inhale the modified air; pulfe 120.

I directed

I directed a warm stimulating plaister to be applied to his side, and five drops of tinctura opii to be given every four hours.—March 28th, pain of his side gone, but his other symptoms continue; bowels regular; has had better nights, but his sleep has been attended with profuse perspirations; the tinctura opii was omitted, and he was directed to take at nearly the same intervals a small glass full of port wine.—April 15th, complaints continue without material alteration. From this date he re-commenced the use of the hydrocarbonate, beginning with it of the strength of one pint to sixteen quarts of common air.—April 25th, at first the modified air occasioned considerable vertigo, but he soon became so much habituated to its operation that the quantity was increased to one, and afterwards to two, quarts. His perspirations have abated, his cough has been less urgent, the quantity of his expectoration has diminished, and the dispnœa, with which he has for some time past been troubled on the slightest motion, is greatly alleviated.

March 3, Since the laſt report he has experienced confiderable amendment, pulfe 98. The quantity of hydrocarbonate was further increafed to a gallon, diluted with four times that quantity of atmofpheric air. May 15th, he has continued to recover fo much in every refpect, that yefterday he was able to walk fourteen miles into the country.

From this time I did not fee him till the middle of June, when he returned to this place with an intention to follow his ufual occupation. He was in every particular fo much better, that he feemed to have recovered his health completely. I advifed him, however, to the contrary, to which he confented, and he has fince been occupied in hay-making, and more lately in reaping. I faw him a few days ago; he cannot be faid either to cough or expectorate, except in the morning, and then in the moſt trifling degree, and his ſtrength is fo completely reſtored, that he has been earning wages equal to thofe of the ſtouteſt of his fellow-labourers,

labourers, with both eafe to himfelf and
fatisfaction to his employer.

CASE V.
Related by Dr. W. Pearfon.

ELIZABETH VYSE, aged 27, having
been feized at the end of autumn with
cough, fever, and fpitting of blood, applied
at the hofpital for relief, and came under
my care laſt October. She informed me
ſhe had been fubject to a cough for three
winters. She had a quick and fmall pulfe,
fluſhed cheeks, dyfpnœa, pain on the fide,
conſtant cough attended with copious expectoration, and night-fweats. She was
very feeble and much emaciated. The hæ-
moptoe was foon removed by the medicines
commonly prefcribed in fuch cafes; but the
fymptoms continued. I therefore ordered
her, on the 12th of November, to breathe
the vapour of vitriolic æther, impregnated
with extract of cicuta two or three times a
day. On the 19th, when I faw her again,
ſhe informed me that ſhe had obtained great
relief

relief from the æther vapour, having much lefs tightnefs acrofs the cheft, and lefs pain of the fide. She faid fhe was fomewhat giddy after every inhalation.—December the 3d, lefs fever, lefs cough, and confiderably better in every refpect. Has found more benefit (to ufe her own words) from the æther application than from any thing elfe.—December 10th, cough and other complaints fo flight, that fhe fays fhe does not require any more medicines.

N. B. During the ufe of æther vapour, fhe took a decoction of Peruvian bark and farfaparilla, and pills compofed of extract of cicuta and rhubarb.

CASE VI.

Related by Mr. Barr. Birmingham, October 9, 1795.

MR. BARROR, of Barton-under-Needwood, being in this town on a vifit to a friend in the fpring of 1793, was feized with an highly inflammatory fever, attended with

with a violent pain of the fide. This fever was followed by a dry tickling cough, a fenfe of tightnefs in breathing, much languor, and a great degree of reftleffnefs and anxiety. His bowels felt full, tenfe, and uneafy; his pulfe intermitted; and he complained that his urine, though nearly in the ufual quantity, did not flow freely, and that he had always the fenfation of not having evacuated the whole. Blifters, boluffes of triturated mercury, and a decoction firft of Peruvian, and afterwards of Anguftra, bark were prefcribed. He was relieved by thefe medicines, but he neither recovered his ftrength nor his fpirits. In this fituation nearly he paffed the remainder of the year in the country; in the fpring of 1794 he came to Birmingham again, with all the fymptoms of his diforder increafed, particularly the oppreffion in breathing. He could neither lie down in bed with comfort, nor afcend the fmalleft aclivity without the greateft uneafinefs. His urine was diminifhed in quantity, and voided with difficulty. A decoction of Seneka root, and fmall small doses of Digitalis were directed and continued for two or three weeks; but they rather seemed to amuse than relieve him. He called on me again last April, and told me that all medicines had lost the power of relieving him, that his breathing was now more generally difficult, that his urine was very scanty, and that his appetite was entirely gone. I prescribed the Digitalis with a itter infusion. He went into the country and continued these medicines for some time. Towards the end of July he called upon me again—but, alas! how changed! His face was now become pale and emaciated, his eyes stared as if taking a last conscious view of their object; which last circumstance much alarmed his friends.— His legs were swelled to such a degree that the skin was become much inflamed, and in danger of bursting. He had a continual tenesmus, and made very little urine; he could not endure an horizontal posture for a moment, but was under the necessity of being bolstered upright in bed through the night; even then he slept little, and that little

little was disturbed and unrefreshing, for he frequently started from his sleep under an impression of instant suffocation.

HAVING seen an account of the happy relief Sir William Chambers had experienced from oxygen in a similar situation, I wrote to my patient, and advised the adoption of the pneumatic plan. 1 did this, I confess, in the present instance, with little hope of advantage; but as the most powerful medicines had produced no salutary effect, I felt it my duty to him, as well as to the cause of humanity, to urge this compliance. I procured him a reading of the case, and the similarity of the circumstances was so striking, that he agreed to place himself immediately under my care.

HE arrived here on the 12th of August, and began to inspire the factitious air on the 13th. I directed one quart of oxygen mixed with nineteen of atmospheric air, to be inhaled every day; but as the symptoms were become extremely urgent, I thought it right to join the use of those active medicines that I had prescribed for him before. Accordingly I directed him to take half a grain of Digitalis in substance, every evening, and four ounces of a decoction of Anguſtra bark in the course of each day. On the third night after inſpiring he found himſelf more compoſed, he could remain longer in one poſture, and the ſtartings, during ſleep, ſeemed both leſs frequent and leſs violent. Every night he was ſenſible of amendment; in ten days he could bear the removal of ſeveral of the pillows that bolſtered him up in bed; and he could ſleep for three or four hours without one ſtarting fit. The ſwelling of his legs too began now to ſubſide; the teneſmus was entirely removed; the quantity of urine was much increaſed, and he could walk up ſtairs with much eaſe; his appetite and cheerfulneſs began to return, and the pale face of diſeaſe to give place to the florid countenance of health. In the courſe of the ſecond week I had gradually increaſed the quantity of oxygen to two quarts a day, diluted as before.

fore. In four weeks from his beginning to infpire the vital air, not a veftige of the diforder remained, except weaknefs; he could lay his head as low in bed as when in perfect health, and fleep the whole night; no fwelling of the legs remained; no difficulty of breathing upon ordinary exertion; and every function was performed with regularity and eafe. He then went home provided with a pneumatic apparatus, and directions how to ufe it, and laid afide the ufe of all medicines except a laxative pill occafionally. He paffed through this town yefterday in perfect health. His ftrength, agility, and vivacity, are greater than in moft men of his age (60).

CASE VII.

Related by Dr. Alderfon.—Hull, June 5th, 1795.

MISS ———, aged 16, had all the fymptoms of approaching phthifis, cold tremors about twelve o'clock; fever, heat, and flufhing

ing every afternoon, pulfe 120, countenance uncommonly florid, breathing rather difficult, cough fevere, accompanied with expectoration; as feveral of her family had died of confumption, there could be little doubt of the tendency of thefe fymptoms; and after finding nitre, fpermaceti, vomits, &c. to have no good effect, I advifed the inhalation of hydrogen air. She therefore daily inhaled a quart of pure hydrogen from water, by every now and then taking an infpiration at the mouth-piece of the tunnel. It frequently occafioned naufea and even vomiting. The pulfe fell, the flufhings and fever fubfided, and the whole train of phthifical fymptoms left her, but at the expence of her fine florid colour, her countenance having ever fince been of a darker tint than before fhe was ill.

CASE VIII.

Related by Mr. Barr.—Birmingham, 14th March 1795.

ABOUT four months ago, a gentleman of this neighbourhood applied to me for advice in the management of a scrophulous ulcer of confiderable extent. He had tried various remedies, but had derived no lasting advantage from any of them. When I first visited him he was worn down by a long course of night-watching. The deep-seated pain of the arm was so constant and severe, that it had in great measure deprived him of sleep. His countenance was pale and sickly; his limbs were continually afflicted with aching pains; every exertion, even the most gentle, seemed beyond the measure of his strength, for his body had lost much of its active power, and his mind much of its wonted energy. The discharge from the ulcer was copious, thin, bloody, and corrosive; and besides, the whole

whole furface of the fore was fo exceedingly irritable, that the mildeft dreffings, applied in the gentleft manner, produced very fevere and lafting pain. During the firft fix weeks of my attendance, he regularly took as much Peruvian bark in fubftance as his ftomach and bowels could bear; and the ulcer was dreffed with various emollient, fedative, and aftringent applications, but without any permanent advantage. I then recommended a trial of oxygen air, which was readily complied with. He began by infpiring four ale quarts diluted with fixteen of atmofpheric air twice a day, and gradually increafed the quantity of oxygen to a cubic foot and a half in the day; by purfuing this plan for about a month, his health was wonderfully improved, but the ulcer fhewed no difpofition to heal. The deep-feated pain was now entirely removed, but in the fpace of a few days more he complained of a burning fenfation over the whole furface of the fore, fimilar to the pain arifing from erifepelatous inflammation. This unpleafant fenfation firft commenced

menced after infpiring the whole quantity of oxygen in the fpace of two hours, which, before had been taken in equally divided portions morning and evening. We ftill purfued our plan, thinking that this new pain might be owing to fome accidental circumftance, and that it would foon pafs away. But it every day continued to increafe, and the ulcer began to fpread wider and wider. The edges became thick, and were turned outwards, and the difcharge became more thin and acrid.

In this fituation a local application feemed proper. I wifhed to have applied hydrocarbonate externally to the ulcer, but this, from fome circumftance of the cafe, was not practicable. I then thought to moderate the ftimulus of the oxygen by a mixture of hydrocarbonate, which Mr. Watt told me would occafion no chemical change in the two airs. Accordingly a mixture of three parts of oxygen, and one of hydrocarbonate, was prefcribed. Four quarts of this mixed air were added to about fixteen of atmofpheric,

ric, and this quantity infpired morning and evening. In lefs than a week the burning fenfation was much diminifhed, and the ulcer put on a more healing appearance. The mixed air was then increafed to five quarts, and ufed as before, which produced an increafe of all the pleafant fymptoms. After a few days trial of this proportion of the mixed air fix quarts were prefcribed. This is the quantity now infpired morning and evening.

My friend, at prefent, enjoys good health and a good appetite, and feels himfelf as ftrong as at any former period of his life. The ulcer is now reduced to lefs than half its original fize, and healing rapidly. There is neither fuperficial nor deep-feated pain remaining, and the action of the contiguous mufcles is free and eafy.

CASE IX.

Related by Dr. Redfearn, Lynn, Norfolk, June 26, 1795.

Mr. B. F—— æt. 23, of a florid complexion, narrow cheft, prominent fhoulders, fmooth fkin, and of a delicate flender form, has been afflicted with hæmoptyfis about two years and a half, attended with dyfpnœa, cough, a difagreeable fenfe of burning in the cheft, and expectoration of a purulent nature. Pulfe about 100, and invariably accelerated by the hydrocarbonate air. The hectic fever was not completely formed, but he had at times a fenfe of chillinefs in the day-time, with heat towards the evening. He began by taking one quart of hydrocarbonate, diluted with twenty-one quarts of atmofpheric air, once a day. From this mixture he experienced much vertigo during its inhalation, and two hours after dinner he fuddenly became vertiginous, from which, however, he foon recovered, although

although a violent head-ach remained during the reſt of the evening.

The following days he only inhaled one pint of hydrocarbonate mixed with twenty quarts of common air, once a day, which generally affected him with ſome ſlight vertigo and tightneſs over his forehead; the hydrocarbonate was increaſed gradually to two quarts or more at one doſe, but I find it always neceſſary to begin with the original doſe, where the air has been recently generated.

My patient has been perſevering in this plan about three months, and has had no return of the hæmorrhage; his cough and expectoration are very much diminiſhed; ſometimes he does not expectorate more than one table-ſpoonful in the ſpace of three days; he has alſo never experienced any of the diſtreſſing heats in his cheſt, which haraſſed him before the adminiſtration of the air; his dyſpnœa is perfectly removed; he can ride upon horſeback twelve miles

miles without feeling much fatigue; his appetite is very good, and he sleeps well; pulse 80; he says he thinks his health is perfectly established.

CASE X.

IN the month of June 1797, a lady began the use of vital air, for an entire loss of voice, which misfortune she had sustained about three years. Her constitution was extremely nervous; had been the greatest part of her life subject to deplorable spasmodic affections, particularly in the organs of respiration, on any trifling exertions of exercise, or in a confined atmosphere; she had been long habituated to an uncommon quantity of opium, to suspend the frequency and violence of her attacks, and many means had been tried in vain to diminish materially the quantity she found necessary for her support.

SHE commenced the use of the vital air under the disadvantage of remaining in

London during the summer, a circumstance which was likely to be attended with great aggravation of her complaints, being contrary to her usual practice, and, in fact, she had already begun to experience extraordinary symptoms of debility from the attempt.

About three quarts of oxygen, with twelve quarts of common air, were administered daily for about a fortnight, the effect of which was, that a slight degree of tightness about her chest came on soon after the inhalation of the air, and generally went off in five or eight minutes time; the pulse was likewise rendered fuller though not more frequent, and the nights were often attended with a sort of restlessness. In consequence of this last effect it was judged proper to diminish the quantity of oxygen air, and accordingly it was found, after repeated augmentations and diminutions of the dose, that about one quart or three pints of oxygen air, with about twelve quarts of common air, was the proportion of the aerial

aerial fluids, which feemed to agree beft with her conftitution. This application was perfifted in for upwards of five months, excepting fome flight intermiffions of a day or two occafionally, and it produced the moft falutary effects. The whole habit of body began to be improved in about a month after the commencement of the application. The fhortnefs of breath vanifhed gradually, as alfo the fymptoms of debility. Her afpect became healthy, and the voice improved by flow degrees, fo that by the end of October its tone was become fully equal to what it ufed to be previoufly to the illnefs. In fhort, this lady now enjoys a better ftate of health than fhe has experienced for many years.

A REMARKABLE circumftance was obferved in this cafe with refpect to the effect of opium, which is, that from long habit fhe had accuftomed herfelf to take an extraordinary quantity of opium daily, in order to fuftain her ufual exertions of the day; for, in fact, the opium produced in her rather a ferenity of fpirits than drowfinefs; but after having inhaled the oxygen air for a few days, fhe found that fhe could do with lefs opium, and in procefs of time fhe further obferved, that the opium, inftead of fupporting, difcompofed her fo much as to oblige her to diminifh the quantity of it to a very fmall part of what fhe had been accuftomed to take before the ufe of the oxygen air.

CASE XI.

Related by Mr. Hey, in a Letter to Dr. Prieftley.

January 8th, MR. LIGHTBOWNE, a young gentleman who lives with me, was feized with a fever, which, after continuing about ten days, began to be attended with thofe fymptoms that indicate a putrefcent ftate of the fluids.

18th. His tongue was black in the morning when I firft vifited him, but the blacknefs blackness went off in the day-time upon drinking; he had begun to doze much the preceding day, and now he took little notice of thofe that were about him; his belly was loofe, and had been fo for fome days; his pulfe beat 110 ftrokes in a minute, and was rather low; he was ordered to take twenty-five grains of Peruvian bark with five of tormentill-root in powder every four hours, and to ufe red wine and water, cold, as his common drink.

19th. I WAS called to vifit him early in the morning, on account of a bleeding at the nofe which had come on; he loft about eight ounces of blood, which was of a loofe texture; the hæmorrhage was fuppreffed, though not without fome difficulty, by means of tents made of foft lint dipped in cold water ftrongly impregnated with tincture of iron, which were introduced within the noftrils quite through to their pofterior apertures, a method which has never yet failed me in like cafes. His tongue was now covered with a thick black pellicle, which was not diminifhed by drinking; his teeth were furred with the fame kind of fordid matter, and even the roof of his mouth and fauces were not free from it; his loofenefs and ftupor continued, and he was almoft inceffantly muttering to himfelf; he took this day a fcruple of the Peruvian bark with ten grains of tormentill every two or three hours; a ftarch clyfter, containing a drachm of the compound powder of bole, without opium, was given morning and evening; a window was fet open in his room, though it was a fevere froft, and the floor was frequently fprinkled with vinegar.

20th, HE continued nearly in the fame ftate; when roufed from his dozing, he generally gave a fenfible anfwer to the queftions afked him, but he immediately relapfed, and repeated his muttering. His fkin was dry and harfh, but without *petechiæ*. He fometimes voided his urine and *fæces* into the bed, but generally had fenfe enough to afk for the bed-pan. As he now naufeated nauseated the bark in substance, it was exchanged for Huxham's tincture, of which he took a table-spoonful every two hours in a cup full of cold water; he drank sometimes a little of the tincture of roses, but his common liquors were red wine and water, or rice-water and brandy acidulated with elixir of vitriol; before drinking he was commonly requested to rinse his mouth with water, to which a little honey and vinegar had been added. His looseness rather increased, and the stools were watery, black, and fœtid; it was judged necessary to moderate this discharge, which seemed to sink him, by mixing a drachm of the *theriaca andromachi*, with each clyster.

21st, THE same putrid symptoms remained, and a *subsultus tendinum* came on; his stools were more fœtid, and remarkably hot; the medicine and clysters were repeated.

REFLECTING upon the disagreeable necessity we seemed to lie under of confining

this

this putrid matter in the inteſtines, left the evacuation ſhould deſtroy the *vis vitæ* before there was time to correct its bad quality, and overcome its bad effects, by the means we were uſing, I conſidered that, if this putrid ferment could be more immediately corrected, a ſtop would probably be put to the flux, which ſeemed to ariſe from, or at leaſt to be increaſed by it, and the *fomes* of the diſeaſe would likewiſe be in a great meaſure removed: I thought nothing was ſo likely to effect this as the introduction of fixed air into the alimentary canal, which, from the experiments of Dr. Macbride, and thoſe you have made ſince his publication, appears to be the moſt powerful corrector of putrefaction hitherto known. I recollected what you had recommended to me as deſerving to be tried in putrid diſeaſes; I mean the injection of this kind of air by way of clyſter, and judged that in the preſent caſe ſuch a method was clearly indicated.

The next morning I mentioned my reflections to Dr. Hird and Dr. Crowther, who

who kindly attended this young gentleman at my requeſt, and propoſed the following method of treatment, which, with their approbation, was immediately entered upon. We firſt gave him five grains of ipecacuanha, to evacuate, in the moſt eaſy manner, part of the putrid *colluvies*; he was then allowed to drink freely of briſk orange-wine, which contained a good deal of fixed air, yet had not loſt its ſweetneſs. The tincture of bark was continued as before, and the water, which he drank along with it, was impregnated with fixed air from the atmoſphere of a large vat of fermenting wort, in the manner I had learned from you. Inſtead of the aſtringent clyſter, air alone was injected, collected from a fermenting mixture of chalk and oil of vitriol: he drank a bottle of orange-wine in the courſe of this day, but refuſed any other liquor, except water and his medicine; two bladders full of fixed air were thrown up in the afternoon.

23d, His stools were less frequent; their heat likewise, and peculiar *fœtor*, were considerably diminished; his muttering was much abated, and the *subsultus tendinum* had left him. Finding that part of the air was rejected when given with a bladder in the usual way, I contrived a method of injecting it which was not so liable to this inconvenience. I took the flexible tube of that instrument which is used for throwing up the fume of tobacco, and tied a small bladder to the end of it that is connected with the box made for receiving the tobacco, which I had previously taken off from the tube; I then put some bits of chalk into a six-ounce phial until it was half filled; upon these I poured such a quantity of oil of vitriol as I though capable of saturating the chalk, and immediately tied the bladder, which I had fixed to the tube, round the neck of the phial; the clyster-pipe, which was fastened to the other end of the tube, was introduced into the *anus* before the oil of vitriol was poured upon the chalk. By this method the air passed gradually into the

the inteftines as it was generated, the rejection of it was in a great meafure prevented, and the inconvenience of keeping the patient uncovered during the operation was avoided.

24th, HE was fo much better that there feemed to be no neceffity for repeating the clyfters; the other means were continued. The window of his room was now kept fhut.

25th, ALL the fymptoms of putrefcency had left him; his tongue and teeth were clean; there remained no unnatural blacknefs or *fœtor* in the ftools, which had now regained their proper confiftence; his dozing and muttering were gone off, and the difagreeable odour of his breath and perfpiration was no longer perceived. He took nourifhment to-day with pleafure, and, in the afternoon, fat up an hour in his chair.

HIS

His fever, however, did not immediately leave him; but this we attributed to his having caught cold from being incautiously uncovered when the window was open, and the weather extremely severe; for a cough, which had troubled him in some degree, from the beginning, increased, and he became likewise very hoarse for several days, his pulse at the same time growing quicker; but these complaints also went off, and he recovered, without any return of the bad symptoms above-mentioned.

Case XII.
Related by Dr. Thomas Percival.

ELIZABETH GRUNDY, aged seventeen, was attacked, on the 10th of December, with the usual symptoms of a continued fever. The common method of cure was pursued, but the disease increased, and soon assumed a putrid type.

On the 23d, I found her in a constant delirium, with a *subsultus tendinum*. Her skin

skin was hot and dry, her tongue black, her thirst immoderate, and stools frequent, extremely offensive, and for the most part involuntary. Her pulse beat 130 strokes in a minute; she dosed much, and was very deaf. I directed wine to be administered freely; a blister to be applied to her back; the *pediluvium* to be used several times in the day, and fixed air to be injected under the form of a clyster every two hours. The next day her stools were less frequent, had lost their foetor, and were no longer discharged involuntarily; her pulse was reduced to 110 strokes in the minute, and her delirium was much abated. Directions were given to repeat the clysters, and to supply the patient liberally with wine. These means were assiduously pursued several days, and the young woman was so recruited by the 28th, that the injections were discontinued. She was now quite rational, and not averse to medicine. A decoction of Peruvian bark was therefore prescribed, by the use of which she speedily recovered her health.

CASE

Case XIII.

Communicated by an intelligent Gentleman in the West of this Island.

A YOUNG lady of 18, the daughter of a neighbour of mine, had been long indiſpoſed with a ſort of diſorder which the medical gentlemen of the neighbourhood could neither properly define, nor in the leaſt relieve. The origin of this indiſpoſition is attributed to a violent cold, which this young lady caught about two years ago at a ball; for ſince that time ſhe had never enjoyed her health, and, in ſpite of all medicines, ſhe rather loſt ground by ſlow degrees than ſhewed any appearance of amendment. The ſymptoms, as nearly as I can deſcribe them, were as follows:

She had loſt all colour from her face and hands, had a remarkably keen appetite, and eat much more than other perſons of her age are wont to do; but this food gave her neither ſtrength nor increment, and ſhe
conſtantly

FACTITIOUS AIRS. 193

constantly complained of wearinefs, refusing to take any sort of exercise. At night she frequently had a slight fever, which terminated by the morning with an head-ach; but this fever did not come on every night, nor did it seem to follow any determinate period. She perspired profusely every night, and even in the day-time the least exertion would throw her into a profuse perspiration. She had tried bark, steel medicines, mineral waters, slight emetics, rhubarb, &c. but all in vain.

In this state of things I first took the liberty of recommending the use of the vital air, or oxygen air, concerning which much has been said of late, and to the physical properties of which I had two years ago been witness in a course of chemical lectures, which I attended in London. After several conversations with the father of the young lady on the subject, it was at last agreed to try the oxygen air, and I undertook to perform the experiment with a few chemical vessels which I happened to have by me.

On the tenth of April, 1797, I put eight ounces of nitre in a small and luted green glass retort, and by exposing the retort to a barely red heat, I obtained nearly two quarts measure of oxygen air, which, being mixed with about eight quarts of common air, was given to the young lady in an awkward manner; for it was introduced into an old glass receiver of an air pump, to the upper aperture of which a leathern tube was adapted, with a glass tube at the extremity of the leathern pipe, to which the patient applied her mouth, &c.

The effect of this application was rather discouraging. The young lady felt a tightness, as she expressed it, about her head, for at least three hours after the inhalation of the vital air, and was very restless at night, in consequence of which she could not be prevailed on to repeat the inhalation for several days. At last, finding that no other bad consequence had been produced by it, she consented to make another trial, which was managed nearly in the same manner; but it was

was attended with much lefs tightnefs about her head, though with an equal degree of reftlefsnefs at night; notwithftanding which a third attempt was made on the following day, and the operation was again repeated after an interval of one day.

ALL thofe trials were more or lefs attended with the like effects as the firft; yet our patient thought that, notwithftanding the reftlefs nights which fhe had paffed, her ftrength feemed to be in fome meafure improved, which encouraged us all to follow the application; and in order to avoid both trouble and expence as much as poffible, we procured one of Mr. Watt's apparatufes from Birmingham, and fome good manganefe from Devonfhire, with which we began to work in a large and more expeditious way.

WE found that Mr. Watt's apparatus requires a nicety of management, without which one may do more harm than good. In the firft trial of this apparatus we got,

inftead

instead of pure air, an explosive elastic fluid; for on lowering a lighted match into a bottle full of it, the air took fire and exploded. It was soon found that this inflammability was occasioned by the moisture which was contained in the manganese, in consequence of which the manganese was made very dry for the subsequent trials, and thus we obtained abundance of oxygen air, which was freed from the carbonic acid air by washing in lime water.

Being now in possession of the proper materials, and having some expectation of success, we began, on the 22d of May, to work assiduously and regularly; and I took care to note, every three or four days, all the circumstances that seemed at all remarkable.

Three pints of oxygen air and eight quarts of common air were administered daily, which constantly produced the tightness of the head and the restlessness at night.

MAY

MAY the 28th, the young lady seemed to have gained strength; but complaining much of the tightness of her head, the quantity of oxygen was diminished to one quart, with eight quarts of common air, per day, which was continued until the 10th of June, by which time her strength was unquestionably improved, and the perspiration at night was considerably diminished; but as a cough happened to come on, we intermitted the application of the vital air for a whole week, after which, the cough having disappeared, the inhalation of the air was recommenced, and continued as before.

BY the beginning of July the good effects of our application were very considerable. The strength of the young lady was such as might be expected in a person of her time of life; the healthy colour was in great measure returned to her face and arms; the perspiration at night was trifling, and she seemed to acquire flesh.

On the 15th of Auguft, the inhalation of the diluted oxygen air was difcontinued, finding that the young lady's health was completely reftored.

Case XIV.

A GENTLEMAN, 35 years of age, of a fcorbutic habit, and fubject to violent head-achs, was induced to try the artificial airs in December 1796, every other medical application having proved ineffectual, and his health gradually declining. He was at firft advifed to try the diluted oxygen air, which he accordingly did, but after three days inhalation of this air, a confiderable degree of inflammation on his lungs obliged him to defift.

The inflammation being fubdued, he again inhaled the oxygen air, and a fimilar effect took place, though this fecond time the inflammation was not fo confiderable.

Finding, therefore, that the oxygen air was not fit for him, he was recommended to drink

drink the water impregnated with carbonic acid gas, and to take some other medicines of a demulcent kind. By following this plan for about six weeks, and by breathing the salubrious air of Devonshire, his health improved to a certain degree; the scorbutic symptoms were reduced, and the head-achs were not quite so frequent as they used to be; but after this improvement, the continuance of the above-mentioned medicines for full three months produced no other effect.

CONSIDERING that in this improved state his constitution might, perhaps, bear the stimulus of the vital air better than it had done before, he was recommended to try that air again, but to take it in smaller quantities. Accordingly he inhaled not more than one pint of it with about sixteen pints of common air every day, which, not producing any inflammation upon his lungs, he continued for upwards of two months, at the end of which time his head-achs were quite vanished, his digestion, which had

had always been defective, was confiderably improved, and he reckoned himfelf quite well.

IN the account of the preceding cafes, the reader may obvioufly remark, that not one unfuccefsful cafe has been introduced; on which it will be proper to mention, that in fo doing I did not mean to imprefs the reader's mind with an exaggerated idea of the power of factitious airs; but that my only meaning was, to render him better acquainted with the practical adminiftration of the aerial fluids, which feemed more likely to be accomplifhed by adducing examples, in which the practice was in fome meafure fanctioned by fuccefs, than otherwife.

WITH refpect to the eftimate of the efficacy of factitious airs in different diforders, my reader muft confult the preceding chapter; for I have expreffed in it the refult, or what feemed to be the fair refult of all

all the cafes that have come to my notice; and of fuch cafes the few that are contained in the preceding pages form a very small part.

I WOULD not be underftood to mean, that the application of the aerial fluids, in the cafes of the prefent chapter, is to be confidered as the model of practical perfection, for in fome of them the adminiftration is evidently incorrect; but they certainly give a great infight into the practice, and hope that, with the affiftance of the cautions and remarks of the following chapter, they may in great meafure prevent the abufe of a new fet of remedies, which have all the appearance of becoming very ufeful tools in the hands of fkilful practitioners.

CHAPTER IX.

PRACTICAL REMARKS, HINTS, &c.

Concerning the Production of Factitious Airs.

IN particular fituations the difficulty of procuring proper materials and proper tools may prevent the poffibility of adopting the moſt expeditious, or, upon the whole, the moſt advantageous, methods of procuring the aerial fluids; and when that is the cafe the practitioner muſt confult the firſt chapter of this effay, for the method which may be more fuitable to the circumſtances of his fituation. But when there is the opportunity of procuring both materials and inſtruments, it is then proper to follow the plan which may appear lefs exceptionable.

THE cheapeſt article for the production of oxygen air is the mineral called *manganefe, nese*, which is found plentifully in many parts of this iſland, and elſewhere. A very good ſort of it is found near Exeter. It ought to be free from extraneous, and particularly noxious, minerals; but it frequently contains a conſiderable proportion of calcareous matter, which may be detected by powdering a little of the mineral, and pouring ſome nitrous acid upon the powder; for this will produce an efferveſcence proportionate to the quantity of calcareous matter. It muſt not, however, be expected to find manganeſe perfectly free from it; for though this may be the caſe with ſmall pieces of that mineral, yet in conſiderably large quantities of it, ſuch as are required for the production of oxygen air, ſome calcareous earth is almoſt always contained; but the only effect which ariſes from it, is the production of carbonic acid gas, together with the oxygen air, the former of which is eaſily ſeparated from the latter by the well-known method of waſhing in lime-water.

THE

The greateſt quantity of oxygen air is extricated from manganeſe merely by the action of a full red heat; it is, therefore, neceſſary to put that mineral in a veſſel capable of reſiſting the action of ſuch a degree of heat. Earthen-ware and certain metals are the materials fit for the conſtruction of ſuch veſſels. The former is certainly unexceptionable in point of purity; but it is not managed very eaſily for this purpoſe, and beſides, the uſe of it is attended with conſiderable expence, for a veſſel of that ſort will hardly ever ſerve more than once, as on cooling after the firſt experiment it generally breaks; and indeed it frequently breaks in the courſe of the experiment. Of the metals, gold or platina veſſels would be the fitteſt for the purpoſe, did not their value offer a material objection. Thoſe metals excepted, iron is the beſt; for though the uſe of a veſſel of this metal be attended with evident objections, yet, when managed with care and attention, the oxygen air may be produced of ſuch a degree of purity as to be

more

more than fufficiently ufeful for medicinal purpofes.

It is neceffary to remark, that in all cafes, but efpecially when an iron veffel is ufed, the manganefe, as well as the veffel in which it is contained, and the pipe or tube which conveys the air from it to the receiver, muft be quite free from animal or vegetable matter, and perfectly dry, otherwife the elaftic fluid, which is produced, may be injured in point of purity, and it may even degenerate into a noxious fluid.

When thofe particulars are attended to, the oxygen air will principally contain a certain proportion of carbonic acid gas, and fome light powder of manganefe, the former of which is to be feparated by means of lime, and the latter will be depofited by ftanding, in about ten or fourteen hours time.

The fpecies of inflammable gas moftly in ufe are extracted by means of diluted

vitriolic

vitriolic acid from zinc or iron, and by paffing the fteam of water over the furface of red hot zinc, or iron, or charcoal.

The gas, extracted by means of diluted acid, holds in fufpenfion fmall particles of the metals concerned, *viz.* of the zinc or the iron, the latter of which in particular may be rendered manifeft by burning the gas in a bottle full of it, in which cafe fome fmall particles of a dark red light will be difcerned within the pale flame of the gas, which are the ferrugineous particles; for thofe minute red fparks are not to be feen in the inflammable gas which is obtained from pond water, or putrid matter, or, in fhort, from fuch fubftances as do not contain any metallic fubftance.

The gas obtained by paffing the fteam of water over red hot zinc, holds in fufpenfion a confiderable quantity of the flowers of zinc, which it depofits in about a day's time.

The

THE gas obtained in a fimilar manner from iron is the moſt abundant, and of courſe the cheapeſt.

FOR the production of the heavy inflammable gas, or hydrocarbonate, Mr. Watt recommends to uſe " charcoal made of the
" twigs of ſofter woods, ſuch as willow,
" poplar, hazle, birch, or ſycamore, avoid-
" ing ſuch as have reſinous or aſtringent
" juices. Prepare the charcoal by heating it
" to full ignition in an open fire, and
" quenching it in clean water, or by filling
" a crucible with it, covering it with clean
" ſand, and expoſing it to a ſtrong heat in
" an air furnace, and then ſuffering it to
" cool. In either of theſe caſes it will be
" found free from any bituminous matter,
" which might contaminate the air, as ge-
" nerally happens with common charcoal."

MR. WATT likewiſe mentions, amongſt other ſorts of inflammable gas, that which is extracted from a mixture of charcoal powder and flaked lime, which, on account

count of its peculiar properties feems likely to prove very ufeful: " In refpect, *fays he,* " to the medicinal properties, all I know is, " that the inflammable air from charcoal " and lime contained no fixed air feparable " by wafhing with quick lime and water, " and that it did not caufe vertigo when in- " haled pure."

ONE or other of thofe fpecies of inflammable gas may be preferred in particular cafes, and it is not only likely, but in great meafure proved by actual experiment, that the particles of iron, or other matter, which are fufpended in a particular fort of gas, may be peculiarly ufeful in certain difeafes.

IN the production of inflammable gas, the introduction of any extraneous matter, and efpecially of vegetable or animal fubftances, and of minerals that contain acids, fhould be carefully guarded againft. It is likewife advifable, for a very obvious reafon, not to conduct this procefs by candle light.

THE

THE carbonic acid gas may be extracted from chalk in Mr. Watt's apparatus, according to the directions given with the said apparatus; but when no extraordinary large quantity of it is required, it is far more commodious to extract it from chalk or marble powder, and diluted vitriolic acid, in a glafs veffel. The difference between chalk and marble in this refpect is, that the former gives out the gas quicker, but is foon exhaufted; whereas the latter gives it out more gradually, and for a greater length of time; hence, in fome cafes, the former, and in others the latter, may be preferred.

Concerning the Prefervation of Aerial Fluids.

OXYGEN Air is not contaminated by keeping in glafs receivers, or in fuch veffels as do not communicate any thing to it, nor does the contact of pure water injure it; but in wooden veffels, or veffels painted with oil paint, and when a confiderable quantity of common river water is in contact with it, the oxygen air will be contaminated more or lefs.

The various fpecies of inflammable gas are apt to degenerate in procefs of time, efpecially if they be kept mixed with common or with oxygen air. The hydrocarbonate, in particular, is vaftly more powerful when frefh made, than two or three days after. Due allowance, therefore, muft be made for the lofs of power in the adminiftration of thofe airs.

When oxygen air, or inflammable gas, is to be taken out of an air-holder or bottle, &c. by putting water in the veffel after the ufual manner, it is advifable to ufe limewater; for the lime will not only abford any carbonic acid gas that may be mixed with thofe airs, but will alfo prevent the putrefaction of the water.

For this purpofe there is no occafion to filtrate the lime-water, as is practifed in the ufual manner of preparing it; but it will be be sufficient to mix the quick lime with the water, and after leaving it at rest for an hour or two, to separate the fluid and useful part from the sediment, by decanting it gently.

The carbonic acid gas is not contaminated by keeping; but as it is absorbed by most fluids, it should not be kept in contact with much water. In most cases it will be better to produce it afresh every time it is wanted.

In order to manage the aerial fluids with the greatest propriety, the practitioner should make himself acquainted with the modes of ascertaining their purity. This may, in great measure, be derived from what has been mentioned towards the beginning of this essay; but if a more particular description of those methods be required, and especially concerning the use of the nitrous-gas eudiometer, or phosphoric eudiometer, the reader must consult those books which have been written expressly on the subject of aerial fluids, as the addition of those methods would increase the bulk of this essay beyond its prescribed limits.

Concerning the Administration of Factitious Airs.

WHEN oil-silk bags or bladders are used, the air or mixture of airs should be introduced immediately before it is to be inhaled, in order to avoid the airs acquiring an unpleasant flavour.

THE oil-silk bags, when not actually in use, should be hung up by means of a string, which may be fastened to the pipe, or they may be put over the back of a chair; but they must not be folded or pressed.

IN the usual way of making the mixture of airs, the factitious gas, in any required quantity, is first introduced in the bag, after which the common air is forced in by means of a common pair of bellows, until the bag is

Factitious Airs.

is quite inflated; for when the capacity of the bag is once known, one may eafily determine the quantity of oxygen or other factitious air, which may be required in order to form a mixture in any given proportion.

When common bellows are ufed for this purpofe, care fhould be had that they be made free from duft and afhes, which are generally contained in fuch bellows as are ufed for common fire-places.

When a perfon is inhaling any fpecies of inflammable gas, or the vapour of ether, the operation fhould be conducted at a diftance from a candle, left the gas fhould catch fire, and at leaft occafion a furprife.

The queftion, which is frequently afked, whether a patient muft be confined to his houfe, or to any particular diet, whilft he is under a courfe of aerial application, fuggefts the propriety of obferving, that there is no particular confinement or diet required merely on that account.

When aerial fluids are adminiftered, it is proper to feel the patient's pulfe both before and after the inhalation, at leaft for the three or four firft inhalations, as thereby one may, in great meafure, be informed of the effect which the aerial fluid is likely to produce, and may regulate the fubfequent applications accordingly.

The patient fhould be enjoined to breathe the mixture of aerial fluids in an eafy and natural way, and not in a forced manner, as fome are apt to do.

With fome perfons the fenfibility of the lungs is fo very great, that they are affected with the fenfation of preternatural heat, and even of inflammation, by a remarkable fmall quantity of oxygen; half a pint of it, mixed with about twenty times that quantity of common air, has been known to produce fuch an effect; and this obfervation has been made where there was not the leaft appearance of miftake, or of any equivocal circumftance. This is particularly the cafe
with

with persons that have recently recovered from an inflammation of the lungs. In such cases, therefore, the practitioner ought to be particularly careful, and he ought to begin by administering very small quantities of oxygen.

THE above-mentioned sensation of heat generally comes on immediately after the inhalation, but sometimes it comes on some hours after, and especially in bed. It is, therefore, necessary to enquire whether any particular effect has been observed at any time between one inhalation and the next, in order to form a proper estimate of the effect of the application.

WHEN this sensation of heat or restlessness is in a trifling degree, the daily inhalation may be continued; but it must he suspended, or at least moderated, whenever it be found to increase by daily repetition.

WHAT has been observed with respect to the effect of oxygen air, may, with proper

per and obvious changes, be alſo applied to the inhalation of other aerial fluids, and particularly of the hydrocarbonate.

However ſtrange and unaccountable ſome of theſe effects may appear, as that produced by a very ſmall quantity of an aerial fluid in certain circumſtances; or that of the preternatural heat coming on ſo long after the inhalation, &c. my reader may reſt aſſured that the facts are true; and though we cannot reconcile the phænomena with the theory, yet as long as abſurdity does not intervene, we muſt not deviate from the path which is pointed out by experience, becauſe we are unable to underſtand the real cauſes of the effects.

APPEN-

APPENDIX.

On the Nature of Blood.

THE intimate connexion between respiration and the state of the blood, the necessary dependance of animal life on the oxygen part of the atmosphere through the intermediation of that fluid, and the various discordant opinions which have been entertained concerning the nature of blood; will easily excuse the introduction of this concise account of the nature of that fluid in the present work, whose principal object is the investigation of the action of aerial fluids on the human body.

THE name of blood has been used in a more or less extended sense by different writers.

writers. Some confine it to the red fluid, which circulates through the veins and the arteries of the animal body; others have, with propriety, extended it to that fluid, which, whether coloured or colourlefs, is the moft abundant in the animal body, and upon the circulation of which the life of the animal principally depends; hence the red colour is not an abfolute characteriftic of blood; and, in fact, the blood of certain animals has not the leaft tint of red. Laftly, the name of blood has been beftowed even upon the fluid which circulates through the veffels of plants.

IN the following pages we fhall extend our obfervations not farther than the red blood, and hardly beyond that of the human fpecies; confining ourfelves principally to the account of facts that are independent on particular opinions.

THIS fluid, fo effentially neceffary to animal life, has been examined with all the ingenuity

ingenuity of man in a mechanical and phyfiological fenſe, as it circulates through the veſſels of the body; it has been carefully viewed, under a variety of circumſtances, through the moſt powerful microſcopes, and it has been analyzed by the moſt ingenious chemiſts. By this means many diſcoveries have been made, and many doubts have been cleared relatively to it; but after all we can form a very inadequate idea of its extenſive uſe and properties. We muſt, however, remain ſatisfied with the preſent knowledge of facts, and muſt leave the farther inveſtigation of the ſubject to the labours and fortune of future obſervers.

BLOOD is a fluid confiſting of a great variety of ingredients, ſome of which are always to be found in it; whilſt others are adventitious, or are to be obſerved in it in particular circumſtances; but the proportion of them all is not only various in different ages, and ſexes, and ſtates of the body, but is not the ſame even in the different

parts

parts of the same body *. A difference not so great with respect to the number, as with respect to the proportion, of the ingredients, has been observed between the blood of man, and that of other animals, such as the ox, the horse, the sheep, the hog, &c. †.

LEAVING it to the physiologists to explain how the blood circulates through the sanguiferous vessels, how the chyle is mixed with it, how a variety of fluids are secreted from it, &c. we shall examine its nature as a fluid out of the body.

BLOOD is of a uniform rich red colour, which inclines towards the florid vermilion

---

* Fourcroy found the blood of a human fœtus to differ in three remarkable particulars from that of an adult; viz. it contains no fibrous substance, strictly speaking, but a sort of gelatinous matter; it does not take a bright colour from the contact of air; and it does not afford any marks of its containing phosphoric acid. The colour of blood is paler and thinner in infants, in women, and in phlegmatic persons, than in men of a healthy and robust constitution.

† Rouelle's Analysis.

in blood that comes out of the arteries, and to the dark purple in blood that comes out of the veins, but the latter, as has been obferved in the preceding pages, becomes brighter by expofure to refpirable air. It is not fo fluid as water, it feels unctuous or faponacious to the touch, and has a little fweetifh or faline tafte.

Soon after its extraction from the body, as the blood cools and remains at reft, a fpontaneous decompofition, or feparation, of parts takes place. A thick lump of coagulated red matter is formed in the middle, called the *craffamentum*, or *clot of blood*, and a fluid of a flight greenifh yellow colour furrounds it, which is called the *ferum*. The quantity of ferum thus formed is lefs at firft, than a few days after; for as the coagulable part contracts and grows harder, fo more and more ferum is forced out of it.

By wafhing the lump of coagulated matter, the colouring fubftance is entirely feparated from it, and the remainder is an infipid,

fipid, tenacious, white, and fibrous, fubftance. The latter is called the *coagulable lymph*, or *fibrous matter*, of the blood. The former, or coloured portion, when viewed through the microfcope, is found to confift entirely of feparate particles, circular and pretty uniform in their fhape; whereas the ferum and the coagulable lymph, when examined with the beft microfcopes, do not appear to contain any diftinct particles in their compofition.

THE blood, therefore, confifts of, or is firft of all refolvable into, three diftinct parts; namely, the *ferum*, the *coagulable lymph*, and the *red particles*; each of which is likewife a compound fubftance, but whofe components cannot be fo eafily feparated from each other.

THE fpecific gravity of human blood is variable, but it always exceeds that of water; the latter being to the former, at the leaft as one to 1,04, and at the moft as one to 1,063. Each of its three principal
components

components is likewife heavier than water; but with refpect to each other, the red particles are the heavieft, and the ferum is the lighteft.

The ferum remains fluid in the ufual temperature of the atmofphere, as far down as a few degrees below the freezing point. But it coagulates in a degree of heat about equal to 160° of Fahrenheit's thermometer. The coagulation of ferum by heat is attended with two peculiar circumftances, *viz.* 1ft, a confiderable quantity of air is extricated from it in the act of congelation; and, 2dly, a fmall part of it does not coagulate, but remains fluid.

The coagulable lymph has been juftly confidered as the moft important part of the blood, and as being the fubftance, from which all the other parts of the animal body derive their increment and their fupport. The fibrous and tenacious nature of this part, which the blood feems to derive from the gluten of our food, is fo remarkable that

that it may be stretched out to a considerable length, and by the continuance of a moderate degree of heat, it may be rendered gradually more and more consistent; so that at last it may be brought to equal the consistency of horn and even of bone.

THE red particles, from which the whole mass of blood derives its colour, seem to have no particular attraction for each other, nor for the other two components, so that in the coagulum they are only entangled and detained by the viscid part. Their peculiar and uniform shape has attracted the attention of philosophers since the latter end of the last century, about which time they were first discovered. They have been attentively examined with the best microscopes, and the appearances which have been partly observed and partly supposed, have given origin to a variety of conjectures and hypotheses, generally fanciful, and often absurd.

WHEN any thin and semitransparent part of a living animal, such as the tale of
a small

a small fish, the membrane which is between the toes of a frog, &c. is viewed through a good microscope; the circulation of the blood through its sanguiferous vessels, is rendered manifest only by the motion of the red particles, which follow each other at a greater or less distance; though in general each particle seems to touch, or, at least, almost to touch the following particle. They never run into each other and incorporate; and though not very hard, they are however possessed of a certain degree of consistency and elasticity; for in passing through small vessels they are frequently seen to assume an elliptical shape, and from other smaller vessels they are absolutely excluded.

Those particles lose their shape, and are dissolved in certain fluids. They are not dissolved in the serous part of the blood, nor in urine, except when they are left in those fluids for some days, or when those fluids are diluted with water. But water is a powerful, and almost an instantaneous solvent of thofe particles; yet water may be deprived of this property by the addition of common falt, or nitre, or of almoft any other neutral falt, as alfo by the admixture of a very fmall proportion of vitriolic acid.

MARINE acid much diluted with water, does not diffolve thofe particles, but it deprives them of their colour.

VINEGAR is likewife a folvent of the red particles, though not fo powerful as water.

WHEN thofe particles have been once dried or diffolved in water, they cannot, by any known method, be made to reaffume their former fhape; and indeed even their formation in the animal body feems to be difficultly accomplifhed, at leaft much lefs expeditioufly than that of the other components of blood; for in perfons that have loft much blood, the fanguiferous veffels are indeed fpeedily filled with new blood; but this blood continues thin and pale for

a con-

FACTITIOUS AIRS. 227

a confiderable time, and if examined through the microfcope, few red particles will be found in it.

UNWILLING to interrupt the account of the chemical properties of blood, I fhall referve the farther examination of the fhape and fize of its red particles for the latter part of this appendix, and fhall now fubjoin the farther analyfis of this fluid, which is principally extracted from Fourcroy's late chemical works.

BLOOD, expofed to a gentle and continued heat, paffes into the ftate of putrid fermentation. When diftilled on a water bath, it affords a phlegm of a faint fmell, which is neither acid nor alkaline, but eafily putrifies, in confequence of its containing an animal fubftance diffolved through it. Expofed to a more intenfe heat, blood gradually coagulates and becomes dry; it then lofes feven-eighths of its weight, and becomes capable of effervefcing with acids. Deficcated blood, expofed to the open air, attracts from it some degree of moisture, and, in the course of a few months, there is formed on it a saline efflorescence, which Rouelle has determined to be carbonate of soda. When distilled by naked fire, it affords a saline phlegm; that is, a phlegm holding in solution an ammoniacal salt, superfaturated with ammoniac. After this phlegm, a light oil passes, then a ponderous coloured oil, and ammoniacal carbonate contaminated with a thick oil. There remains in the retort a spungy coal, very difficult to be incinerated, which is found to contain muriate of soda, carbonate of soda, oxyde of iron, and a matter apparently earthy, which seems to be calcareous phosphate.

Blood, when burnt in a crucible, affords several products, in the following order: 1. water, and a little ammoniac; 2. oil, and carbonate of ammoniac, which forms a yellowish vapour, thicker than the former; 3. Prussic acid, which is easily distinguished by its fœtid smell of peach-flowers; 4. phosphoric

phoric acid, which is formed by the combustion of phosphorus, and is not disengaged till the blood be reduced to a coal; 5. carbonate of soda, which is volatilized at an intense heat; 6. after this there remains in the crucible only a blackish, granulated, cryftallized oxyde of iron, mixed with calcareous phofphate. The ferrugineous particles of this laft product may be feparated by the magnet, especially when the refiduum has been previoufly heated together with charcoal-powder in a covered crucible.

BLOOD combined with alkalis, without previous decompofition, becomes more fluid by ftanding. Acids inftantaneoufly coagulate it, and alter its colour. By filtrating this fubftance, evaporating the liquor paffed through the filter, drying it before a moderate fire, and lixiviating the matter that has been dried, neutral falts are obtained, confifting of foda, with the acid that was mixed with the blood.

IF entire blood, mixed with a fourth part of its weight of water, be coagulated by heat, and if a part of the fluid that fwims above the coagulated portion be evaporated, a fubftance of a brown yellow is obtained, which is eafily diftinguifhed to be true bile.

THE ferum, which has been lately called the *albuminous fluid*, communicates a green tinge to fyrup of violets. By diftillation on a water-bath, it affords a phlegm of a mild infipid tafte, which is neither acid nor alkaline, but fpeedily putrifies. After lofing this phlegm, it is dry, hard, and tranfparent like horn : it is no longer foluble in water: by diftillation in a retort, it affords an alkaline phlegm, a confiderable quantity of ammoniacal carbonate, and a very fœtid thick oil. All thefe products, in general, have a peculiar fœtid fmell. The coal of the ferum, when diftilled by naked fire, almoft entirely fills the retort. It is fo difficult to incinerate, that it muft be kept burning for feveral hours, and expofed to a great

great deal of fresh air, before it can be reduced to ashes. The ashes are of a blackish grey colour, and contain muriate and carbonate of soda, with calcareous phosphate.

The serum, if exposed for some time to an hot temperature in an open vessel, passes readily into a state of putrefaction, and then affords a considerable quantity of ammoniacal carbonate, with an oil, the smell of which is insufferably nauseous.

This liquor combines with water in any proportion, and then it loses its consistency, its taste, its greenish colour. When poured into boiling water, coagulates, almost wholly, instantaneously. A portion of this fluid forms, with the water, a sort of opaque and milky white liquor; which, according to Bucquet, possesses all the characteristic properties of milk, *viz.* it is rarified, and caused to mount up, by heat, and is coagulated by the same agents, *viz.* by acids, and by alcohol.

The serum possesses the property of fixing and rendering solid by heat, two or three times its weight of water. But when the water exceeds seven times the quantity of serum, then no coagulation takes place.

Alkalies render the serum more fluid, and acids coagulate it. This last mixture, filtrated and evaporated after filtration, affords a neutral salt formed of soda and the acid employed; which shews that soda exists in the serum in a naked state, in full possession of all its properties. The coagulum formed in this liquor by the addition of an acid, is very speedily dissolved in ammoniac, which is the general solvent of the albuminous part of the blood; but it is not dissoluble at all in pure water. Acids precipitate this matter in union with ammoniac. The coagulum affords, by distillation, the same products as the serum desiccated, and its carbonaceous residue contains a good deal of carbonate of soda.

THE ſerum, inſpiſſated, affords azotic gas by the action of the nitric acid, with the help of a moderate heat. On increaſing the fire, there is a quantity of nitrous gas diſengaged from the mixture.

THE ſerum does not decompoſe calcareous or aluminous neutral ſalts; but it acts with ſufficient energy in decompoſing metallic ſalts.

THIS fluid is liable to be congealed by alcohol; but this coagulum differs from that formed by means of acids, chiefly for its ſolubility in water.

THE ſerum, therefore, appears to be an animal mucilage, conſiſting of water, acidifiable oily baſes, muriate and carbonate of ſoda, with calcareous phoſphate.

THE clot of the blood affords, by expoſure to the heat of a water-bath, an inſipid water; it becomes, at the ſame time, dry and brittle. It affords, in the retort, an alkaline

kaline phlegm, a thick oil, of a fœtid, empyreumatic smell, and a good deal of ammoniacal carbonate. The residuum which it leaves, is a spongy coal, of a sparkling metallic aspect, difficult to incinerate, and affording, when treated with sulphuric acid, sulphate of soda and sulphate of iron; there remains, after these operations, a mixture of calcareous phosphate with carbonaceous matter. When exposed to a hot atmosphere, the clot of blood readily putrifies.

When the clot is divided, by washing, into its two principal components, *viz.* the red part which is dissolved in the water, and the coagulable lymph; if the former be treated with different menstrua it will be found possessed of the same characteristics with the serum; excepting that it contains a greater proportion of iron. The latter, after being well washed, will remain white, colourless, and insipid. It affords, by distillation on a water-bath, an insipid phlegm, without smell, and liable to putrefaction. Even the gentlest heat hardens this fibrous matter

matter in a singular manner. When exposed suddenly to a strong fire it shrinks like parchment. By distillation, in a retort, it affords an ammoniacal phlegm, a ponderous oil, which is thick and very fœtid, and a good deal of ammoniacal carbonate, contaminated with a portion of oil. The residual coal is not very bulky, but compact, ponderous, and easier incinerated than that of the serum. Its ashes are very white; it contains no saline matter, as it must have lost, by the washing, whatever is contained of that kind; and no iron; it is a sort of residue of an earthy appearance, and seemingly calcareous phosphate.

THE fibrous part of the blood putrifies very quickly and easily. When exposed to a hot moist atmosphere, it swells, and affords a good deal of ammoniac. It is not soluble in water; when boiled in that fluid, it becomes hard, and acquires a grey colour. Alkalies do not dissolve it, but even the weakest acids combine with it. The nitric acid disengages from it a considerable
quantity

quantity of azotic gas, and of Pruffic acid, which comes out in vapour, and at length diffolves it with effervefcence, and the difengagement of nitrous gas. When it ceafes to emit nitrous gas, the refidue is obferved to contain oily and faline flakes, fwimming in a yellowifh liquor: this liquor affords, by evaporation, oxalic acid in cryftals; and at the fame time, depofits no inconfiderable quantity of flakes, compofed of a peculiar oil, and calcareous phofphate. It appears, that hydrogene, carbone, and azote, which conftitute the fibrous fubftance, are feparated in different proportions, to combine with the oxygen of the nitric acid, and thus form the Pruffic and carbonic acids that are difengaged in gas, and the oxalic and malic acids, that remain in folution, and are feparated only by cryftallization,

The fibrous matter diffolves alfo in the muriatic acid, which converts it into a fort of green jelly. The acid of vinegar diffolves it with the help of heat: water, and efpecially alkalies, precipitate this fibrous
matter

matter from acids. This animal matter is decompofed in thefe combinations; and when feparated, by whatever means, from acids, no longer exhibits the fame properties.

Thus much may fuffice with refpect to the chemical properties of blood. I fhall now return to the examination of the configuration of its red particles, with which I fhall conclude this effay.

The red particles, which form a very fmall part of the human blood, were difcovered by means of the microfcope, towards the end of the laft century. They were found to be circular and uniform; a tranfparent flat furface appearing to be furrounded by a dark circumference. This peculiar fhape feemed to indicate their being of fingular ufe to the animal œconomy, and excited the induftry of philofophers to the further inveftigations of their ftructure. As this could only be obtained without ufing more perfect microfcopes, and as the perfection of microfcopes depended

on

on the construction of small lenses, various methods were contrived for the attainment of this object, and microscopical lenses of very short focuses, and of course of great magnifying power, were soon produced; but the utmost power of those lenses could only discover that when the particles of blood were magnified beyond a certain number of times, they exhibited a dark speck in their middle, as a center to their circumference.

This is all that could be clearly discerned in those particles by means of ground lenses; but a vast deal more was suggested by the imagination; and it is curious to observe how much the eye and the understanding were deceived by the natural imperfection of the instruments, and by the influence of premature theories.

Finding that the improvement of ground lenses, beyond the abovementioned power, was obstructed by weighty practical difficulties, the deficiency was attempted to be supplied

supplied by the use of globules of glass made by melting; for in the state of fusion, the natural attraction between its particles, will easily form the glass into a spherical body. Several methods were accordingly devised for constructing those globules, as may be seen in Dr. Smith's Optics, and other publications; but those methods are either defective or absolutely impracticable. And, in fact, I do not find that any globules of very great magnifying power, were used before the time of *Father della Torre*, who, about the middle of the present century, constructed globules of wonderful minuteness, and at the same time clear and distinct.

THIS Neapolitan Friar, who, without much scientific knowledge, possessed a considerable share of ingenuity, made many observations with those magnifiers, which he published, together with a minute and faithful account of his method of constructing the glass globules, in a pamphlet, about 30 years ago. But both the construction

of those globules, and their use as magnifiers, are very difficult; so that few persons have attempted to repeat Torre's experiments, and amongst those, fewer still have been successful. This want of success has thrown a considerable degree of suspicion on Torre's observations; and as few people are liberal enough to acknowledge their want of sufficient patience and address, the failure of the attempts has generally induced people to consider Torre's assertions in the light of mistakes or exaggerations. " The " Abbé Torre," *says a recent writer*, " ex-
" amined the red particles of blood with
" simple lenses too; but they magnified so
" highly, that from this cause all his noisy
" mistake has arisen; for he used not ground
" lenses, but small sphericles of glass, form-
" ed by dropping melted glass into water:
" they magnified so much that to him the
" central spot appeared much darker; he
" said that these were not globules, but
" rings. He sent his sphericles of glass,
" and his observations from Italy, his own
" country, to the Royal Society; and for a
" long

" long while, though nobody could fee
" them, ftill the public were annoyed by
" Abbé Torre's rings *."

SOME years ago, when Torre's publication firft became known to me, I endeavoured to conftruct microfcopical globules after his method, and to repeat his obfervations. The undertaking, which at firft fight appeared clear and eafy, proved on trial very difficult and laborious; however, after perfevering for a confiderable time, I at laft procured three or four ufeful globules out of a vaft number of imperfect ones. With thefe globules, and an apparatus made exprefsly for fuch delicate experiments, I repeated feveral of Torre's obfervations, and (as far as I now recollect, for both the globules and the journal of obfervations have been long fince loft) I found that his defcription of appearances is very accurate, though his conjectures may fometimes be crude or miftaken.

* Bell's Anatomy, vol. ii. p. 89.

Being lately intent on the subject of the present work, I was desirous to repeat the above-mentioned microscopical observations, and for this purpose I obtained, after a considerable expenditure of time and labour, a few glass globules, sufficiently useful, and with them I made the observations which I shall now lay before the public. But it will be proper to premise a concise account of the principal opinions that have been entertained by various ingenious persons, concerning the construction of the red particles, as the origin of some of those opinions will be evidently pointed out by the observations that follow.

Leeuwenhoeck thought that each red particle of the blood consisted of, and was resolvible into, six smaller globules, and that every one of these secondary globules consisted of other smaller particles. Hewson took them for bladders which contained a nucleus or central body that seemed to roll from one side of the bag to the other.

Torre

Torre faw them like rings; *viz.* confifting of an internal and an external circle, and this ring appeared to be divided, or to confift of parts joined together like the rim of a common coach wheel. Falconer confidered them as flat or fpheroidical bodies; for he thought he fometimes faw them fideways. " The red globules," *fays the late Mr. J. Hunter*, " are always nearly of the
" fame fize in the fame animal, and when
" in the ferum do not run into one another
" as oil does when divided into fmall glo-
" bules in water. This form, therefore,
" does not arife fimply from their not unit-
" ing with the ferum, but they have really
" a determined fhape and fize. This is
" fimilar to what is obferved of the globules
" in milk; for milk being oily, its globules
" are not foluble in water; neither do they
" confift of fuch pure oil as to run into
" each other; nor will they diffolve in oil.
" I fufpect, therefore, that they are regular
" bodies, fo that two of them could not
" unite and form one *." Dr. Wells is of

* Treatife on the Blood, p. 41.

opinion, that the red globules confift of two parts, one within the other, and that the outer, being infoluble in ferum or dilute folutions of neutral falts, defends the inner from the action of thofe fluids *.

Much having been faid againſt the ufe of microfcopical glafs globules, efpecially by perfons who had never feen them, I thought it neceffary to afcertain the limits of the fuppofed diftortion of the image, and other imperfections that had been attributed to them, and for this purpofe I viewed certain objects of fimple or determinate figures through lenfes and globules of different powers, increafing gradually from a magnifying power of about eight or ten as far as that of about 400 times in lineal extenfion.

A DELICATE ftraight line made by means of a diamond on a piece of glafs, and which was quite invifible to the naked

* Phil. Tranſ. P. II. for 1797.

eye,

eye, when thus gradually magnified, appeared always ftraight, provided it was made to pafs through the axis of the lens or globule. The feathers of a butterfly, or rather any particular part of one of thofe feathers, never changed its figure though magnified upwards of 400 times.

THERE is an evident diftortion of the image when the object is viewed through the edge of the lens, and efpecially of a globule; but no perfon verfed in fuch experiments will ever obferve through the edge of lenfes, though the lenfes be ever fo perfect.

WHEN the object is not very flat, it is then evident that a perfect view of it can not be had at once; for if one part of it be in the focus, the reft of the object muft of courfe be out of it; yet by alternately bringing one part of the object and then another to the focus, one may, in moft cafes, acquire a fufficiently accurate idea of its fhape.

Various obfervations of this fort gave me reafon to conclude that the glafs globules are by no means fo imperfect as they have been reprefented. Their diftortion of the image is trifling and limited; the tranfparency of fome of them (and fuch only fhould be ufed) is equal to that of the beft polifhed lenfes; but the ufe of thofe globules is very difficult, and it is on account of this difficulty that they have been neglected and mifreprefented.

For the fake of thofe who may be willing to repeat fuch experiments, I fhall barely mention the principal difficulties which attend the ufe of the globules.

Their focus is confiderably nearer the furface than that of a lens of the fame magnifying power; and as a globule, in order to magnify more than the ufual microfcopical lenfes, muft be lefs than the 30th of an inch in diameter, and its focus fhorter than the hundredth part of an inch, it follows that the common microfcopical apparatufes are

are in general inapplicable to fuch globules, fince the deviation of one or two thoufandth parts of an inch in the adjuftment of the focus will occafion a confiderable degree of indiftinction.

It is for the fame reafon, that the globule muft be fet fo as to have part of its furface actually out of the brafs cell, and yet it muft be fecured fo as not to drop out.

The brafs cell muft admit of the globule being eafily taken out and replaced; for when they are obfcured by duft, &c. to which they are very fubject, they can feldom be cleaned without removing them from the cell.—Let us now return to the particles of blood.

I have repeatedly meafured the diameters of the red particles, both by means of my mother-of-pearl micrometer in a compound microfcope, and likewife by looking at them with one eye through a fingle lens,

lens, and referring their image to a scale properly divided, and viewed with the other eye out of the microscope.

In persons of nearly the same age the mean size of the particles differs very little indeed. In the same person they differ a little, and their figure is not very circular. This deviation from the circle is not such as a flat circular surface would assume in its different inclinations to the axis of vision; for, according to the rules of orthographic projection, the flat circular surface must appear either circular, or elliptical, or as a straight line; whereas I never saw the particles of blood as straight lines, *viz.* edgewise, and the elliptical figure, which they sometimes assume, is by no means regular.

In an adult of the human species, the diameters of the red particles run from about 0,0003 to about 0,0004 parts of an inch, and I very seldom saw one smaller or larger than those limits. If, therefore, we take

Factitious Airs.

take the smallest particles and set them in a row, we shall find that about 3334 of them will equal one inch, and if we take the largest, about 2500 of them will measure one inch.

When the particles are magnified more than 40 or 50 times, and less than 80 (meaning always in diameter), they appear like colourless transparent spots inclosed within dark circles.

When magnified more than 80 times, and less than about 160, a dark spot, like a dot made with ink on paper, appears in the middle of each particle.

If the reflector which illumines the particles, instead of being situated straight before the object, be set on one side of the axis of vision, so as to throw the light obliquely on the object, then the half of the dark circle of each particle disappears, *viz.* that half which is on the side opposite to the reflector.

reflector. The central spot does at the same time appear to change its place.

When the particles are magnified above 200 times, the central spot appears converted into a circle inclosing a transparent space. The diameter of this inner circle is about the half of that of the external one; but the proportion of these diameters, or the size of the internal circle, may be caused to increase or decrease by the least alteration of the distance between the object and the microscopical lens; and by the same means the space within the inner circle may be rendered clearer or darker than that between the two circles. The position of the inner circle is changed by the direction of the light; for if the particle of blood be viewed through a microscopical globule, directly facing the flame of a candle, without the intermediation of any lens or reflector, the inner circle will appear concentric with the outer one; but if the candle be moved a little to one side, so that the light may fall obliquely on the particle

of

FACTITIOUS AIRS. 251

of blood, then the inner circle will be observed to move towards the oppofite fide, and to acquire an elliptical fhape.

When the particles of blood are magnified above 400 times, an imperfect image of the candle, which is placed before the microfcope, may be feen within the inner circle of each particle.

Through a glafs globule of 0,018 of an inch in diameter, I have feen the red particles of blood magnified about 900 times, in which cafe the image of the flame of the candle could be feen within the inner circle of each particle very clearly, at leaft fo as to fhew to which fide the motion of the air in the room inclined it.

Notwithstanding this great magnifying power, the annulus or fpace between the two circles did not appear to be divided, excepting fome accidental fractures, which now and then could be feen in a few of the particles.

These

These obfervations feem to prove, that the red particles of blood are not perforated, but that they are globular, and of fome uniform fubftance much lefs tranfparent than glafs. They likewife fhew that Mr. Hewfon's idea of their containing a central body or nucleus, moveable within the external fhell, arofe from the apparent change of place which the various direction of the light produces on the central fpot or inner circle of each particle. Warned, however, by the example of other obfervers, I fhall not attempt to offer any farther conjectures concerning the nature and conftruction of thofe particles. My reader may draw what conclufion he thinks proper from the above-mentioned facts, and he may alfo, with little trouble, fatisfy his curiofity concerning thofe appearances, as I find that microfcopical glafs globules may be had at Mr. Shutleworth's philofophical inftrument fhop on Ludgate Hill. I fhall therefore conclude with the account of a few experiments which I have made, with a view of imitating the phenomena that are exhibited

by

by the particles of blood, the refult of which feems to corroborate what has been already obferved.

On the fuppofition of the red particles being globular, I expected that globules of other tranfparent matter would exhibit the fame appearances as the particles of blood, and my expectations were in great meafure verified by actual experiments.

A GLASS globule was placed as an object upon the ftage of the microfcope, and was fucceffively viewed through lenfes of various, but not great, magnifying powers. As every part of the globule could not be at once in the focus, the whole of it was not of courfe equally diftinct. This indiftinction, however, being not very great, I fhall proceed without taking any farther notice of it.

The globule appeared like a dark circular furface, with a tranfparent circular fpot in the middle, and in this fpot there appeared

peared a diſtinct image of the candle or the window, or, in ſhort, of any other object that was placed directly before it.

In this experiment three points of difference between the glaſs globule and the particles of blood were remarked, *viz.* 1ſt, that the globule ſhewed a diſtinct, whereas the particle ſhewed an indiſtinct image of the candle; 2dly, that the inner circle of the globule is much ſmaller in compariſon with its external boundary, than the inner circle of the particle is in compariſon with its external one; and, 3dly, that the annulus or ſpace between the two circles is uniformly dark in the glaſs globule, whereas in the particle it is about as clear as the internal ſurface, or rather clearer.

The firſt and the laſt of theſe points of difference ſeem to depend on the imperfect tranſparency of the particles of blood; for in ſemitranſparent bodies, whatever light falls upon any part of them is ſcattered through the whole body.

The

THE second point of difference I attributed to the particles of blood being surrounded by a coagulated fluid of nearly an equal refracting power with themselves, whereas the glass globule was surrounded by air only. In order to verify this supposition, I placed the glass globule in water, and viewing it in that state through the same magnifiers that had been used before, I found that the transparent part or circle appeared much larger than in the former case *.

IN the globule of glass, as well as in the particle of blood, the inner circle may be made to appear larger or smaller, by altering

---

* These appearances are perfectly reconcileable to the doctrine of optics. The light, which falls from a luminous object upon the glass globule, illuminates at most one half of its surface, and in entering the surface of the glass, it is refracted towards the axis of the globule; hence the whole cone of light being contracted, must pass through a small part only of the opposite surface, and must leave the rest destitute of light. Now this contraction of the light must vary according to the difference between the refractive power of the globule and that of the surrounding medium.

the diftance between the object and the microfcopical lens.

IN the glafs globule the inner circle may be feen to move from the middle of the dark furface, according as the candle is moved from the direct line between the object and the microfcopical lens.

A:

AERIAL fluids, different species of, *page* 1—breathed in combinations of three or four at a time, 49—injected into the cellular membrane, 51—abforbed by water, 56—theory of, 58—their conftituent principles, 67—how applied to the human body, 91, &c. 114—remarks concerning the production of, 202—prefervation of, 209—remarks concerning the adminiftration of, 212.

Acidulous foda water, 145.

Air, common or atmofpherical, its properties, 2—its purity afcertained by means of nitrous gas, 3—contains a variety of fubftances, 5—vitiated various ways, 6—methods of meliorating it, 6—refpiration of, 23—imbibed through the pores of the fkin, 51—injected into the cellular membrane, 52—its conftituent principles, 67—decompofed in the lungs, 72—its various qualities, 88.

Animal heat, origin of, 82.

Animation, fufpended, treatment of, 116.

S *Apparatus*,

I N D E X.

Apparatus, neceſſary for the production of factitious airs, 90—for adminiſtering the factitious airs, 91.

Aſthma, treatment of, 118.

Azotic gas, what, 6—deleterious to animals, 48—injected into the cellular membrane, 52—injected into the jugular vein, 53—its constituent principles, 67—a diluent of oxygen air, 87.

B.

Bag, oil ſilk, for containing air, &c. 92—muſt not be folded, &c. 212—how to correct its bad ſmell, 93.

Bladders, for containing air, 94—how to correct their bad ſmell, 95.

Blood, its colour altered by the preſence of different aerial fluids, 53, 77—imbibes the oxygen in the lungs, 72, 76, 81—its colour in the courſe of circulation, 78, 83, 220—name of, 217—conſiſts of various ingredients, 219—decompoſition of, 221—its ſpecific gravity, 222.

C.

Cancer, treatment of, 120.

Calculous complaints, treatment of, 144.

Caloric, what, 62—laws relative to the communication of, 63—capacity of bodies for containing it, 64—produces a total change of properties in bodies, 66.

Capacity, for containing caloric, different in different bodies, 64—proportion of, in different bodies, 65.

Carbonic acid gas, its properties, 15—whence obtained, 16—reſpiration of, 46—appearances in the bodies of animals

INDEX.

mals that have died in it, 47—injected into the cellular membrane, 52—injected into the jugular vein, 53—its pungency, 55—an antiseptic, 55, 102, 111—how administered, 56, 97—its constituent principles, 68—produced in the process of respiration, 73, 85.

Cases, medical, 149.

Catarrh, treatment of, 123.

Chlorosis, treatment of, 123.

Combustion, theory of, 69—analogous to respiration, 75.

Consumption, treatment of, 124.

Coughs, treatment of, 130.

Crassamentum, or clot of blood, what, 221.

D.

Debility, treatment of, 130.

Dephlogisticated air, the same as oxygen air, 1.

Digestion, impaired, treatment of, 133.

Dropsy, treatment of, 135.

Dyspepsia, treatment of, 133.

E.

Eudiometer, what, 3.

Eruptions, treatment of, 136.

Ether, its use, 110.

F.

Fevers, treatment of, 137.

Fixed air, the same as carbonic acid gas, 1.

INDEX.

H.

Head-ach, treatment of, 141.
Heat, its origin, 62, 66—animal, whence derived, 82.
Hæmoptysis, treatment of, 142.
Hydrocarbonate, what, 19—respiration of, 45—remarks concerning the production of, 207.
Hydrogen gas, its properties, 17—whence obtained, 18—respiration of, 42, 44—properties of its various species, 43—injected into the cellular membrane, 52—it does neither accelerate nor retard putrefaction, 55—its constituent principles, 67—remarks concerning the production of, 205, 208—contains extraneous particles, 206—apt to degenerate, 210.

I.

Inflammable air, the same as hydrogen gas, 2.
Inflammation, of the lungs, how treated, 107, 109, 118.

L.

Lime-water, use of, 210.
Lungs, human, affected by the smallest differences in the purity of the air, 4, 214.
Lymph of blood, 222, 223.

M.

Manganese yields abundance of oxygen air, 14, 202—it generally contains calcareous earth, 203—should be made quite dry before it is used, 205.
Mouth-piece, used for breathing any particular sort of air, 96.

N. *Nitrous*

INDEX.

N.

Nitrous gas, diminishes common air, 2—diminishes oxygen air, 10.

O.

Oil-silk bags, useful for respiring particular airs, 92—how to remove their oily smell, 93.
Ophthalmia, treatment of, 142.
Oxygen air, its properties, 9—obtained from various substances, 9, 11—diminished in various circumstances, 10, 14—respiration of, 29—its stimulant quality, 34, 35, 36, 37, 54, 101—effects produced by it when diluted with common and other airs, 34, 38, 40—imbibed through the pores of the skin, 51—injected into the cellular membrane, 52—its constituent principles, 67—indispensably necessary to animal life, 74—when to be administered, 102, 104.
Oxygenation, of metals, 69—of blood, 72.

P.

Paralysis, treatment of, 143.
Phlogisticated air, what, 6.
Phlogiston, what, 60.
Phosphuret of hydrogen, what, 19.
Phthisis, treatment of, 124.

R.

Red particles of blood, 222, 224—opinions concerning their construction, 242—microscopical appearance of, 249—their diameter, 248.
Respiration of common air, 23—of oxygen air, 29—of other factitious airs, 40—theory of, 58, 71.

S.

Scurvy, treatment of, 143.
Serum of blood, 221, 223.

INDEX.

Skin, the aerial fluids are imbibed and expelled through its pores, 51.
Specific gravities of the aerial fluids, 20—table of, 21—effects arising from it, 49.
Stone in the bladder, treatment of, 144.
Sulphuret of hydrogen, 19.
Swellings, treatment of, 146.

T.

Theory, of aerial fluids, 58—of respiration, 59, 60, 72.

U.

Vessels, proper, for the production of aerial fluids, 204—for the preservation, &c. 209.
Ulcers, treatment of, 147.

W.

Water absorbs aerial fluids, 56—its constituent principles, 68—expelled from the lungs, 73—origin of, in the lungs, 85—how impregnated with carbonic acid gas, 98
White swelling, case of, 146.

THE END.

Works publiſhed by the ſame Author.

I. A TREATISE on the Nature and Properties of AIR, and other permanently elaſtic Fluids; to which is prefixed, an Introduction to Chemiſtry. Quarto, with Plates.

II. A TREATISE on ELECTRICITY, in Theory and Practice, with original Experiments. The Fourth Edition, in Three Vols. Octavo, with Plates.

III. The HISTORY and Practice of AEROSTATION, Octavo, with Plates.

IV. A TREATISE on MAGNETISM, in Theory and Practice; with original Experiments. The Second Edition, Octavo, with Plates.

V. Two MINERALOGICAL TABLES, with an Explanation and Index.

VI. DESCRIPTION and USE of the Telescopical Mother-of-Pearl MICROMETER, invented by the Author.

☞ The following Directions are added for the Information of such Readers as may be willing to try the Factitious Airs, or may otherwise wish to promote the Subject of this Essay:

Mr. WATT's Apparatus for the Production, &c. of FACTITIOUS AIRS, is sold at Chippindall's Birmingham Warehouse, in Salisbury-Court, Fleet-Street.

The ARTIFICIAL AIRS, accurately prepared, are sold in any Quantity by J. Rance, No. 30, Clipstone-Street, Fitzroy-Square.

The ACIDULOUS SODA WATER, containing an extraordinary Quantity of Carbonic Acid Gas, is sold at Schweppe's Artificial Mineral Waters Manufactory, No. 11, Margaret-Street, Cavendish-Square.

The MICROSCOPICAL GLASS GLOBULES, such as are mentioned in the Appendix, are sold by H. Shuttleworth, Optician, on Ludgate-Hill.

www.ingramcontent.com/pod-product-compliance
Lightning Source LLC
Chambersburg PA
CBHW031949230426
43672CB00010B/2103